ARTIFICIAL INTELLIGENCE: ITS SCOPE AND LIMITS

STUDIES IN COGNITIVE SYSTEMS

VOLUME 4

The titles published in this series are listed at the end of this volume.

ARTIFICIAL INTELLIGENCE: ITS SCOPE AND LIMITS

JAMES H. FETZER

Department of Philosophy,
University of Minnesota, Duluth, U.S.A

KLUWER ACADEMIC PUBLISHERS

DORDRECHT / BOSTON / LONDON

ISBN 0-7923-0505-1 (HB)

Published by Kluwer Academic Publishers,
P.O. Box 17, 3300 AA Dordrecht, The Netherlands.

Kluwer Academic Publishers incorporates the publishing programmes of
D. Reidel, Martinus Nijhoff, Dr W. Junk and MTP Press.

Sold and distributed in the U.S.A. and Canada
by Kluwer Academic Publishers,
101 Philip Drive, Norwell, MA 02061, U.S.A.

In all other countries, sold and distributed
by Kluwer Academic Publishers Group,
P.O. Box 322, 3300 AH Dordrecht, The Netherlands.

Printed on acid-free paper

Printed in the Netherlands

To
Henry W. Davis

TABLE OF CONTENTS

SERIES PREFACE

This series will include monographs and collections of studies devoted to the investigation and exploration of knowledge, information, and data-processing systems of all kinds, no matter whether human, (other) animal, or machine. Its scope is intended to span the full range of interests from classical problems in the philosophy of mind and philosophical psychology through issues in cognitive psychology and sociobiology (concerning the mental capabilities of other species) to ideas related to artificial intelligence and to computer science. While primary emphasis will be placed upon theoretical, conceptual, and epistemological aspects of these problems and domains, empirical, experimental, and methodological studies will also appear from time to time.

The perspective that prevails in artificial intelligence today suggests that the theory of computability defines the boundaries of the nature of thought, precisely because all thinking is computational. This paradigm draws its inspiration from the symbol-system hypothesis of Newell and Simon and finds its culmination in the computational conception of language and mentality. The "standard conception" represented by these views is subjected to a thorough and sustained critique in the pages of this book. Employing a distinction between systems for which signs are significant for the *users of* a system and others for which signs are significant for *use by* a system, I have sought to define the boundaries of what AI, in principle, may be expected to achieve. The point of view elaborated here explains the reasons why AI can succeed, even though there grounds for doubting that computers will ever think.

J. H. F.

FOREWORD

> (A) cognitive theory seeks to connect the *intensional* properties of mental states with their *causal* properties vis-à-vis behavior. Which is, of course, exactly what a theory of the mind ought to do.
>
> Jerry Fodor (1980)

In spite of the vast literature that has appeared since the Turing Test was introduced in 1950, the most fascinating question about artificial intelligence remains whether or not machines can have minds. Obviously, it is not the case that just *any* machine might have a mind: even if your lawn mower doesn't always want to start, that does not suggest (to most of us, at least) that it has a mind of its own. No, the paradigm cases – those that are most likely to have minds if *any* machines have minds – appear to be a special class of digital machines. The problem, of course, is figuring out exactly what it is about these specific things by virtue of which we should (or should not) ascribe to them the mental powers of human beings. And the question is literal rather than metaphorical, for we want to know if things like these possess the same – or similar – mental powers as human beings.

The rationale for thinking so derives its major motivation from a plausible analogy between human beings and these digital machines. What we know about them both is that they exhibit certain forms of overt behavior under certain conditions. Human beings tend to respond in certain ways to specific stimuli, mediated by internal processes. Digital machines likewise produce certain sorts of output when given certain input, when equipped with a suitable program. It is tempting to infer that the foundation for an analogy – which might be said to be "The Basic Model" behind AI – lies here:

	Human Beings:	*Digital Machines*:
Domain:	Stimuli	Inputs
Function:	Processes	Programs
Range:	Responses	Outputs

The Basic Model

Thus, the mental processes of human beings are viewed as functions from stimuli to responses just as the computer programs of digital machines are viewed as functions from inputs to outputs. Their general features are assumed to be the very same, which, if true, would yield enormous benefits.

The cornucopia of rewards that The Basic Model promises to deliver is very great, indeed. That this is the case can be discerned from the perspective of the philosophy of mind. For there are three great problems within the philosophy of mind – the nature of mind, the relation of mind to body, and the existence of other minds – and The Basic Model might solve them all. The problem of the nature of mind, of course, concerns the properties, characteristics, and features that properly qualify as "minds". If The Basic Model is correct, then they are the same properties, characteristics, and features that collectively constitute programs. The relation of mind to body, from this point of view, is simply that of "software" to "hardware". And the existence of other minds can be suitably resolved by the Turing Test.

Most of these benefits (not necessarily all) come as a package deal. The mental processes of human beings are like the computer programs of digital machines, after all, *only if* at some appropriate level human beings function as – or, as John Haugeland (1981) has put it, "simply *are*" – digital machines. The relation of software to hardware, moreover, illuminates the relation of minds to bodies *only if* the mental processes of human beings function as (or "simply *are*") the computer programs of digital machines. And the existence of other minds can be resolved by the Turing Test *only if* the test succeeds as a usually reliable evidential indicator of the presence of mentality itself. Thus, if the Turing Test were flawed for its intended purpose, it might turn out to be the case that some other test could serve its function better; but it is difficult to see how The Basic Model could otherwise withstand alteration.

My objective in this work is to lay out the evidence for believing that The Basic Model is wrong. I shall argue that human beings and digital machines are causal systems but of very different kinds; that a distinction has to be drawn between symbol and semiotic systems; that digital machines are symbol systems, while humans are semiotic; that only semiotic systems are possessors of minds; and that the prevalent conceptions of AI are based upon mistakes. If the arguments that I offer are correct, The Basic Model is not merely wrong in its specific details but also wrong in its general conception. The Turing Test does not resolve the problem of the existence of other minds. The relation of software to hardware does not provide a solution to the mind/body problem. Mental processes are not computer programs, and human beings are not digital machines. Even the conception of "programs" as functions from inputs to outputs cannot withstand critical investigation.

The net result, however, is not therefore negative, for AI can succeed in spite of these findings by embracing a more modest and rationally warranted conception of its boundaries. The representation of knowledge; the development of expert systems; the simulation and modeling of natural and behavioral phenomena; and the rapid processing of vast quantities of data, information and knowledge are destined to assume a major role in shaping the future of our species. And just as the Agricultural Revolution relieved us of a great deal of hunting and gathering and the Industrial Revolution relieved us of a great deal of physical labor, the Computer Revolution will relieve us of a great deal of mental effort. A more adequate conception of the scope and limits of AI confirms that, within the confines of what is possible, AI can provide methods and tools of immense benefit to all mankind.

4 July 1989 James H. Fetzer

ACKNOWLEDGMENTS

The author is grateful to the following publishers for permission to incorporate material from the following articles: "Signs and Minds: An Introduction to the Theory of Semiotic Systems", in *Aspects of Artificial Intelligence*, J. H. Fetzer, ed., Dordrecht, The Netherlands, Kluwer Academic Publishers, 1988, pp. 133-161; "Language and Mentality: Computational, Representational, and Dispositional Conceptions", *Behaviorism* 17 (1989), pp. 21-39, Cambridge Center for Behavioral Studies; and "Program Verification: The Very Idea", *Communications of the ACM* 31 (1988), pp. 1048-1063, The Association for Computing Machinery.

My interest in computer science generally and in artificial intelligence specifically was greatly enhanced through a post-doctoral fellowship spent at Wright State University in Dayton, Ohio, during 1986-87. Among those with whom I studied were Henry Davis, David Hemmendinger, Jack Kulas, and Al Sanders. I have also benefitted from discussions with Charles Dunlop, Michael Losonsky, Terry Rankin, William J. Rapaport, and David Smillie as well as from suggestions that were proposed by the anonymous referees.

It is a pleasure to acknowledge the financial assistance provided by the Graduate School of the University of Minnesota in the form of a Summer Faculty Research Fellowship, which enabled me to bring this project to its completion. Special assistance was provided by Nancy Olsen with respect to the preparation of the figures and diagrams. And I am grateful to Derek Middleton, my acquisition editor, and to others on the staff of Kluwer Academic Publishers for the support and assistance they have provided over the years.

J. H. F.

PART I

METAMENTALITY

1. WHAT IS ARTIFICIAL INTELLIGENCE?

One of the fascinating aspects of the field of artificial intelligence (AI) is that the precise nature of its subject matter turns out to be surprisingly difficult to define. The problem, of course, has two parts, since securing an adequate grasp of the nature of the artificial would do only as long as we were already in possession of a suitable understanding of the idea of intelligence. What is supposed to be "artificial" about artificial intelligence, no doubt, has to do with its origins and mode of creation in arising as a product of human contrivance and ingenuity rather than as a result of natural (especially biological or evolutionary) influence. Things that are *artificially intelligent*, in other words, differ from those that are *naturally intelligent* as artifacts that possess special properties ordinarily possessed by non-artifacts. So these are things that have a certain property (intelligence) as a result of a certain process (because they were created, designed, or manufactured in this way).

These artifacts, of course, are of a certain special kind in the sense that they are commonly thought of as being *machines*. If machines are simply things that are capable of performing work, then, since human beings are capable of performing work, human beings turn out to be machines, too. If anyone wants to know whether or not there are such things as intelligent machines, therefore, the answer seems obvious so long as human beings are (at least, some of the time) *intelligent*. A more difficult question, whose answer is less evident, however, arises from distinguishing between animate and inanimate machines. Human beings, after all, may result from human contrivance and ingenuity, but they are clearly biological in their origin. The issue thus becomes whether or not *inanimate* machines, as opposed to human beings, are capable of possessing a certain special property that human beings are supposed to display – if not always, at least on certain special occasions.

The problem revolves about the identification or the definition of what is meant to be "intelligent" about artificial intelligence. The dictionary may seem to be an appropriate place to begin. *Webster's New World Dictionary* (1988), for example, defines "intelligence" as "a) the ability to learn or understand from experience; ability to acquire and retain knowledge; mental ability; b) the ability to respond quickly and successfully to a new situation; use of the faculty of reason in solving problems, directing conduct, etc., effectively; c) in *psychology*, measured success in using these abilities to perform certain tasks". If the concept of intelligence is to be applicable (at least, in principle) to inanimate machines, however, it must not be the case that inanimate machines could not possibly be intelligent as "a matter of definition".

Matters of definition, of course, are often thought to be arbitrary or capricious, capable of resolution merely by mutual agreement: "The question is who is to be master, that's all!" But much more hinges upon the meaning that we assign to the words that we use when we address specific problems and search for suitable answers. In this case, the establishment of an appropriate definition for "intelligence" becomes equivalent to a theory of the nature of intelligence as a phenomenon encountered in the world. Hence, for example, any analysis that has the effect of *excluding* human beings from the class of intelligent things, on the one hand, or that has the effect of *including* tables and chairs within the class of intelligent things, on the other, would thereby display its own inadequacy. By reflecting upon "exemplars" of the property in question – instances in which its presence or absence and other features these things possess are not in doubt – it should be possible to achieve conceptual clarification, if not to secure complete agreement, on the crucial features of intelligence itself, which amounts to a prototype theory.

Even if we assume that human beings *are* among the things that happen to be intelligent, the problem remains of isolating those specific aspects of human existence that are supposed to be "intelligent". Since human beings exhibit anger, jealousy, and rage, it might be asked if inanimate machines would be intelligent if they could exhibit anger, jealousy, and rage. Notice, the point is not the trivial one that we are likely to mistake anger, jealousy, and rage for "intelligence" as a matter of fact, but rather the subtle one that *unless we already know that anger, jealousy, and rage are not aspects of "intelligence"*, we are not able to rule them out. If human beings are the best exemplars of the property in question, yet are widely observed to exhibit anger, jealousy, and rage, then by what standard, yardstick, or criterion can we tell that these are *not* some of the features characteristic of intelligence? And, in fact, there are situations, circumstances, and conditions under which anger, jealousy and rage would be intelligent. So how can we possibly tell?

While these definitions may seem helpful with respect to human beings, they remain problematic in application to machines. If we take for granted that human beings do possess the ability to learn or understand from experience, for example, that in turn invites elucidation of precisely what happens when something – it could be anything at all – "learns" or "understands" from experience. So unpacking "intelligence" now becomes unpacking "learning or understanding from experience". This may appear to be a doubtful advance, especially since whether or not inanimate machines can learn or understand from experience appears to be no less troublesome than the question whether or not inanimate machines ever have the capacity to be intelligent. Similar reflections obtain with respect to the ability to acquire and to retain knowledge and with respect to mental ability in general: whether inanimate machines can acquire and retain knowledge (or whether inanimate machines can have minds) appears to offer few benefits in understanding "intelligence" – unless we already understand the nature of knowledge or of learning or of mentality, in which case theoretical progress might yet be possible, after all.

CRITERIA OF INTELLIGENCE

If we possessed the practical ability to sort things out, to determine of any specific thing whether or not it ought to be included within the class of intelligent things, of course, it might not matter if we were to be less than completely successful in devising a theory of intelligence. In this case, perhaps a *criterion* could serve in lieu of a *definition*, where the criterion functions as a (usually reliable, but not therefore infallible) evidential indicator for deciding, in a given case, whether or not that case is an instance of this property. Indeed, several tests of this general type have been proposed, among which the most famous is that devised by Alan Turing, which is now known as "the Turing Test". If the Turing Test – or another criterion – were to prove to be a usually reliable evidential indicator of the presence or the absence of intelligence, that might be enough (even if it were not infallible).

The Turing Test. Turing (1950) offered a suggestion that has turned out to have widespread appeal, especially because it offers the hope of resolving these difficult problems at a single stroke. The test he proposed is a special case of what is known as "the imitation game", in which two contestants are pitted against one another, but in opposite ways, with respect to some third party. An example might be the separation of a male and a female behind a curtain, where the third party attempts to guess the sex of the contestants, based strictly upon their answers to questions he poses. The female might attempt to lead the third party into the realization that she is female, while the male would endeavor to deceive him into believing that he is female as well. The means of communication between parties, of course, would have to be one that would not reveal the sex of the communicant for the game to be a success. What is most important is that almost any property might do.

Turing recognized that an especially interesting application of the imitation game would pit an inanimate machine against a human being, where the property in question was not their sex but their intelligence. He reasoned that if, under the appropriate conditions (including, therefore, a suitable means of communication, such as by teletype machine), a third party could distinguish between the human being and the inanimate machine no better than he could distinguish between a male and a female on the basis of their answers to questions he might pose, as before, then it would be reasonable to conclude that the parties were equal with regard to the property in question. The criterion that might be used to empirically ascertain whether or not a machine possesses intelligence, in other words, would be its capacity to fool the third party – a human being – into believing that it is human, too. The result would depend upon its success in inducing a specific false belief.

The Chinese Room. Indeed, from its conception around 1950, the Turing Test remained largely unchallenged as a suitable criterion for machine intelligence until confronted by the scenario that has come to be known as "the Chinese Room", posed by John Searle [(1980) and (1984)]. Searle offered his argument as a counterexample to certain claims about *machine understanding* that were then being advanced by Roger Schank on behalf of "scripts" (with fascinating features that we shall explore in Chapter 7). But the same case tends to undermine the plausibility of the Turing Test with respect to machine intelligence, precisely because it tends to undermine the plausibility of the imitation game within which it assumes its significance. The Searle counterexample thus provides an indirect argument against the Turing Test, but one that – as we shall discover – can receive considerable reinforcement.

The argument Searle fashioned is a thought experiment of the following kind. Imagine that someone (Searle himself, in particular) is locked into an enclosed room with one entry through which Chinese symbols are now and then sent into the room and one exit through which (the same or different) Chinese symbols could be sent out. Thus, if the occupant of that room, who is fluent in English but ignorant of Chinese, had in his possession a book of instructions, written in English, directing that certain specific Chinese characters should be sent out when certain other Chinese characters happen to be sent in, then if that person were to act in accordance with those instructions, it might appear to those outside the room as though, contrary to the hypothesis, he understands Chinese. While this special arrangement might fool observers into believing he understands Chinese, however, he does not.

Indeed, Searle strengthened his argument by suggesting that the characters sent into the room might be called "input", the characters out "output", and the book of instructions "a program"; yet however impressive its performance might appear, it would not be the case that an input-program-output system of this kind actually understood Chinese. Even though its publicly observable inputs and outputs might be indistinguishable from those of another input-program-output system that actually understood Chinese (say, the publicly observable inputs and outputs of a fluent Chinese interpreter), any inference to the conclusion that this system actually understood Chinese would be completely mistaken, since that conclusion would be untrue.

Looking back upon the imitation game itself from this point of view, moreover, it is striking to realize that, even in the first instance, if a male were successful in convincing a third party that he were female, that obviously would not change his sex! Analogously, if an input-program-output system of the kind Searle envisions were to fool outside observers into the belief that it understood Chinese, that obviously would not mean that this system actually understood Chinese! And – most importantly – if an input-program-output system were to display behavior that was indistinguishable from the behavior displayed by a human being, that would not dictate that that system really possessed intelligence! There is, after all, a fundamental difference between appearance and reality. The net impact of Searle's case thus appears to be a drastic reduction in the plausibility of the Turing Test.

The Korean Room. Some students of AI have drawn different conclusions from the premises described. William J. Rapaport, especially, takes a different tack, suggesting that the Turing Test can be entertained as a tacit reflection of suitable conditions for the possession of intelligence. Rapaport seeks to dispurse the damage wrought by the Chinese Room argument with a variation of his own, which is intended to demonstrate that

Searle's argument may be less conclusive than it initially appears. He proposes the scenario of a Korean professor who is an authority on Shakespeare, even though he does not know the English language. Having studied Shakespeare's work in excellent Korean translations, he has overcome what for others appeared to be an insuperable obstacle to master the plays. His articles on the Bard, which have been translated into English on his behalf, have met with great acclaim. He is viewed as an expert scholar [Rapaport (1988), pp. 114-115].

The point of this fanciful account is that the Korean professor stands to Shakespeare's plays as the-man-in-the-room stands to Chinese. Yet, as Rapaport himself insists, it would surely be a mistaken inference to conclude that the Korean professor does not understand *something;* and, by parallel reasoning, it would similarly be a mistaken inference to conclude that the-man-in-the-room understands *nothing*. The difficulty encountered at this juncture, of course, is unpacking precisely what it is that this Korean professor is supposed to understand, since even Rapaport does not deny that, whatever it may be, he certainly does not understand *English*. By parallel reasoning, once again, however, it seems to follow that for the-man-in-the-room, whatever he is supposed to understand, it is certainly not *Chinese*.

Even more important, from a theoretical point of view, however, is the perspective Rapaport brings to bear upon the Turing Test itself. Correctly observing that Turing (1950) rejected the question, "Can machines think?", in favor of the more behavioristic question, "Can a machine convince a human to believe that it (the computer) is a human?", Rapaport infers that,

> To be able to do that, the computer must be able to understand natural language. So, understanding natural language is a necessary condition for passing the Turing Test, and to that extent, at least, it is a mark of intelligence. I think, by the way, that it is also a sufficient condition. [Rapaport (1988), p. 83]

Whether or not the capacity to understand natural language is a necessary condition for the possession of intelligence, of course, invites controversy in the case of non-human animals, many of which display behavioral patterns that certainly *seem* to be intelligent. (Ask any owner of a cat!) But even if understanding natural language is merely a sufficient condition, rather than both necessary and sufficient for intelligence, it raises intriguing questions.

Rapaport could argue, for example, that the appearance-of-understanding qualifies as understanding with regard to understanding the plays, even though the appearance-of-being-of-female-sex does not qualify as being-of-female-sex. The issue might then become whether this really is an instance in which appearance *is* reality [a position that Rapaport (1988) and Shapiro and Rapaport (1988) have proposed]. Such a defense, however, although ingenious, possesses very little *prima facie* plausibility. That it might appear to outsiders as though the-man-in-the-Korean-room understands Shakespeare's plays, after all, is no more contrary to the hypothesis that he does *not* understand Shakespeare's plays than that it might appear to outsiders as though the-man-in-the-Chinese-room understands Chinese is contary to the hypothesis that he does *not* understand Chinese. Hence, as a criterion of intelligence, the Turing Test still seems to be an unreliable or a poor one.

Notice that, although Rapaport views his conception as essentially the same as Turing's, there are several reasons for doubt. Turing's Test, in particular, concerns whether or not a machine *could fool a human into thinking that it is human too*. So its success hinges upon whether or not a belief could be induced by the behavior of a machine without concern for whether or not that machine actually understands anything at all. Rapaport's Test, by comparison, concerns whether or not a machine *could really understand natural language*. Rapaport's Test depends upon whether or not a machine actually understands a natural language without concern for whether or not it might use its linguistic ability to induce beliefs in any human being. The Turing criterion concerns belief states and the conditions under which they might justifiably be acquired, while Rapaport's concerns the kind of thing a machine happens to be apart from any beliefs that we might form about it.

While Rapaport wants to interpret his test as a criterion of intelligence as-detected-by-the-Turing-Test itself, moreover, the benefits that may be derived from its introduction tend to depend on presuming that the Turing Test functions as a criterion of intelligence, whereas Rapaport's Test serves as something akin to a definition. Notice, especially, that if the ability to understand natural language were both necessary and sufficient to possess intelligence, then anything that were unable to understand natural language could not be intelligent, while anything that were intelligent could not fail to understand natural language. And this remains the case whether Rapaport intended it as a clarification of the Turing Test or not. Consequently, I shall interpret Rapaport's Test as proposing a definition of the property under consideration rather than merely positing a criterion of its presence.

VARIETIES OF INTELLIGENCE

There are at least three reasons for appreciating the point of view that Rapaport's Test supplies. The first concerns the focus it provides upon the kind of thing that something happens to be rather than the beliefs that we fallible human beings might form about it. Questions about what kinds of things there happen to be fall within the domain of *ontology* and are often referred to as "ontic" issues. Questions about the beliefs that we, as fallible human beings, happen to adopt, by contrast, fall either within the domain of *psychology* and are frequently referred to as "cognitive" matters (when they concern the beliefs we do adopt) or within the domain of *epistemology* and are referred to as "epistemic" questions (when they concern the beliefs we ought to adopt). The differences between them are immense: Turing's Test is an epistemic criterion, for example, while Rapaport's is an ontic definition instead.

If Rapaport's Test takes us from a criterion to a possible definition of intelligence, it also has the virtue of focusing attention upon the pivotal role of language. It is widely assumed that there is a fundamental relationship between language and thought, especially when the suggestion is proposed that all thinking takes place in language. Without language, there could be no thought, if this suggestion is correct. Thus, by accenting the importance of language with respect to the nature of intelligence, such a test invites us to consider an ability that may very plausibly be supposed to be far more essential to the nature of intelligence than anything like the ability to fool a human being into embracing a false belief. And, indeed, whether or not Rapaport's Test ultimately survives as a definition of intelligence, we shall discover ample grounds for regarding the ability to understand natural language as a crucial factor in distinguishing natural and artificial intelligence.

The third reason for appreciating Rapaport's Test is much more subtle, reflecting as it does the importance of implied *success* in such descriptions as "the ability *to learn and understand* from experience, ability *to acquire and retain* knowledge, ability *to respond quickly and successfully* to new situations", and the like. All of these characterizations, including Rapaport's conception of the ability *to understand* natural language, imply success in undertaking a certain activity or in exercising a specific faculty or ability. Notice, in particular, that *attempting* to learn, *trying* to acquire and retain, *wanting* to respond quickly – but *without* success – would not ordinarily be entertained as exemplifications of intelligence. Indeed, from this point of view, there is an important difference between "things that can do things successfully" and "things that can do things by using their minds". There are lots of things, after all, that can do things by using their minds, where those things are stupid, foolish, or otherwise lacking in intellectual merits.

Notice that an automated mechanism, say, a device for capping bottles as they emerge from a production line, might perform its intended function quite successfully and very dependably without raising any questions as to whether or not it possessed intelligence. In fact, all sides – every party – to these disputes would be inclined to agree that this machine (call it a robot, if you will) does not possess any intelligence at all. Yet the successful way in which it performs its task would certainly merit the accolade implied by reference to this entity as an "intelligent machine". No doubt, there is room for debate, even in simple cases like this one, depending upon the features of mechanisms of this kind. Nevertheless, later on, when we have occasion to reflect upon what we have discovered in the meanwhile, it may turn out to be worthwhile to consider just what examples such as this have to tell us.

Rationality. Taking a cue from the last of these issues, yet another way the problem of intelligence might be addressed would be as a synonym for rationality. The idea of rationality, however, is highly ambiguous, insofar as there appear to be at least three varieties of rationality, which, in turn, suggests the possibility that there might be three corresponding varieties of intelligence. The first of these concerns *the rationality of ends*, meaning the rationality of choosing or selecting specific goals, aims, or objectives as worthy of pursuit. While we often take it for granted that anyone can try to do anything they want – whether or not it is illegal, immoral, or fattening – if someone seriously, rather than wishfully, wanted to discover a number which is both even and odd (where these terms mean what they ordinarily mean within the theory of numbers), then anyone aware of this state of affairs might think something was amiss, precisely because it is logically impossible, i.e., strictly inconsistent, for any number to be both odd and even.

Hence, a necessary condition for the rationality of ends appears to be that the attainment of that aim, objective, or goal must not be a logical impossibility. That this requirement is clearly not sufficient follows from reflection upon another class of cases, including individuals who sincerely rather than wistfully want to be in two places at the same time, to consume food without exercising and still lose weight, etc. (How familiar they look!) In these cases, what is wrong is not that the situations described are logically inconsistent; these are logically possible states of affairs that might have been possible if only the world had been different in certain respects. For these scenarios are fanciful precisely because their realization could occur only by violating natural laws (concerning the locations of physical things, the relations between food consumption and weight loss, etc.). It appears as though a rational goal must be a physical as well as a logical possibility.

Suppose, however, that someone wanted to secure a state of affairs whose attainment was neither logically nor physically impossible, perhaps by being the first man to climb Mt. Everest, the second man to marry Elizabeth Taylor, or something such. In cases of this kind, the pursuit of these goals would be pointless and without purpose, not because it would be logically impossible (in relation to a particular language) or because it would be physically impossible (in relation to the world's own laws), but because it would be historically impossible (in relation to the history of the world up to now). One cannot do (for the first time) something that has already been done, even though whether or not it can be done (for the first time) is a function of the historical past rather than of logic or of physical laws.

There are other types of rationality, of course, including, in particular, *rationality of action* and *rationality of belief* [cf. Hempel (1962)]. Rationality of action involves choosing means that are appropriate to attain one's ends, where means are "appropriate" in relation to one's ethics, abilities, capabilities, and opportunities. It would be irrational in this sense for a skinny vegetarian with no other source of support to enter a hamburger eating contest for a $5.00 prize because he aspires to travel to Europe on the Concorde. Rationality of belief, by comparison, involves accepting all and only those beliefs that are adequately supported by the available relevant evidence, where the form this evidence could take might be perceptual, inductive, or deductive, as we shall subsequently ascertain. It would be irrational in this sense for someone who has seen (direct and indirect) evidence that man has landed on the moon to fail to believe that man has landed on the moon, unless the total available evidence were to override that belief. These are matters we shall discuss again in Chapters 4 and 5.

Intentionality. On the basis of these reflections, the evidence tends to suggest that rationality (in any of its senses) does not provide a promising avenue toward understanding "intelligence" in the sense appropriate to AI. But the reason this is so should not be overlooked, since it represents (what appears to be) one of the fundamental differences between human beings and inanimate machines. For rationality of ends, rationality of action, and rationality of belief tacity presuppose the existence of agents, organisms, or entities whose behavior results (at least in part) from the causal interplay of motives and beliefs, where *motives* are wants and desires of a system with *beliefs* that might possibly be true. But these conditions are not very likely to be ones that an inanimate machine should be expected to satisfy.

Notice that the rationality of ends, for example, involves choosing or selecting appropriate goals, aims, or objectives as worthy of pursuit, where the propriety of those choices from a subjective point of view is a function of the range of available alternatives, the means that might be employed in their pursuit, and the morality of adopting such means to attain those ends. One view of the most defensible conception of decision-making activities by rational human beings involves envisioning them in terms of preference relations between alternative states of affairs (or "payoffs"), which represent – explicitly or implicitly – expected utilities (in some appropriate sense) as a function of expectations and desirabilities. [Eells (1982) reflects some recent philosophical work in this area, while Cohen and Perrault (1979), Allen and Perrault (1980), and Halpern (1986) exemplify some approaches in AI.]

Important theoretical questions arise here between different accounts of decision-making, including the respective merits of optimizing, satisficing, and cost-benefit decision policies. For our purpose, however, what is significant about all this is that these are extremely implausible questions to ask in relation to any machine. One can always inquire as to the principles of design that entered into a machine's construction, of course, but it is simply silly to raise questions such as whether the bottle-capping robot described above would prefer to do something else instead. Even if we surprisingly frequently indulge ourselves by ascribing human properties to inanimate things – suggesting, for example, that our lawn mower might not want to start (as though it were driven by motives and desires) – we accept these practices as metaphorical vestiges of anthropomorphic tendencies of the past. They are nothing more than a very convenient manner of speaking.

As though he could convert vice into virtue, Daniel Dennett (1971) has endorsed "the intentional stance", which occurs when beliefs and desires *are* ascribed to inanimate things, such as chess-playing machines. Thus,

> Lingering doubts about whether the chess-playing computer *really* has beliefs and desires are misplaced; for the definition of intentional systems I have given does not say that intentional systems *really* have beliefs and desires, but that one can explain and predict their behavior by *ascribing* beliefs and desires to them.... The decision to adopt the strategy is pragmatic, and is not intrinsically right or wrong. [Dennett (1971), pp. 224-225]

The principal advantage of adopting the intentional stance, in Dennett's estimation, is that "it is much easier to decide whether a machine can be an intentional system than it is to decide whether a machine can *really* think, or be conscious, or morally responsible" [Dennett (1971), p. 235]. On that score, there can be no doubt. But the fundamental questions still remain.

It should be obvious that the intentional stance, at best, offers (what we may think of as) a practical technique to adopt in attempting to explain and to predict the behavior of inanimate things. In emphasizing the pragmatic dividends of embracing such an attitude, however, Dennett implicitly endorses the position known as *instrumentalism*, which maintains that we need not believe in the existence of entities and properties that lie beyond our observational capacity or our perceptual range but may treat them instead as, say, "convenient fictions". An alternative position, known as *realism*, denies that imperceptible and unobservable entities and properties are therefore any the less real. While the evidence for their existence must be indirect in character, when it supports theories that posit their existence, it supports belief in their existence as well. Dennett's account appears to be a special case of instrumentalism regarding the existence of mental entities.

Some questions, of course, are hard to answer. In suggesting that the intentional stance poses a question that is easier to answer than "whether a machine can *really* think", Dennett has changed the subject. We are not looking for an answer to the question, "Can it be beneficial to regard some inanimate machines as though they were thinking things?" The answer to that question, I am willing to agree, is almost certainly, "Yes". Even here, however, the benefits are less vast than Dennett promises them to be. If machines really lack beliefs and desires, "explanations" ascribing to them properties that they do not have surely cannot be adequate – they cannot even be true! So long as truth is a condition for the adequacy of an explanation, the most that the intentional stance can provide is predictive utility.

Moreover, it should also be apparent that Dennett has taken two steps backward in embracing an epistemic criterion in lieu of an ontic definition. If our objective is to understand whether or not machines can have minds, his position does not help us. But it can be useful to come to the realization of how widespread the practice has become of taking for granted a certain mode of speech whose consequences we would tend to deny, if only they were made explicit. And it calls back to mind the Turing Test itself insofar as it hinges upon the question of whether or not some digital machine might be able to fool a human being into believing that it is human too. For surely it makes sense to talk about a machine "fooling" a human being only if that effect has been achieved deliberately (or on purpose), i.e., as a causal consequence of that machine's own motives and beliefs. Otherwise, the best that can be said is that a human might mistake the behavior of this machine for that of a human being but not that he has been "deceived" by such a thing.

Humanity. A more candid and less question-begging response to this entire problem complex has been advanced by Eugene Charniak and Drew McDermott. In their impressive survey of AI, Charniak and McDermott suggest, "Artificial intelligence is the study of mental faculties through the use of computational models" [Charniak and McDermott (1985), p. 6]. The assumption underlying this conception is that, at some suitable level, the way in which the brain functions is the same as the way in which certain computational systems – digital machines, in particular – also function. Yet they also maintain, "The ultimate goal of AI research (which we are very far from achieving) is to build a person, or, more humbly, an animal" [Charniak and McDermott (1985), p. 7]. Science fiction addicts might rejoice and theologians may cringe, but there, stated as baldly as could be, is what to many appears to be a fantasy: AI is attempting to create an artificial human being!

For the moment, let us defer consideration of the theoretical tenability of this conception and instead attempt to appreciate the various ingredients whose availability would be required to make its realization even remotely plausible. Specifically, the situation can be schematized in terms of a simple stimulus-process-response framework or model, where stimuli are "inputs", responses are "outputs", and processes – for now – may be viewed as "black boxes". The precise content of these black boxes is absolutely crucial to AI as Charniak and McDermott describe it, since if, for example, what is going on inside a black box when it generates outputs from inputs is not a computational process, then AI is – to that extent, at least – a failure. If the reason the black box is not a computational process is that AI research has not yet captured the way in which humans operate, then it may just be a failure in practice; but if AI research has captured the way in which humans operate and the process is not computational, then it has to be a failure in principle.

Hence, the following diagram suggests the model that tends to be tacitly assumed within AI when human beings are compared with digital machines:

	Human Beings:	*Digital Machines*:
Domain:	Stimuli	Inputs
Function:	Processes	Programs
Range:	Responses	Outputs

Fig. 1. The Basic Model.

Notice, in order for this analogy to be correct (or well-founded), it has to be the case that human beings can be characterized by processes that are properly viewed as functions from (the domain) stimuli to (the range) responses and that digital machines can be characterized as well by programs that are properly viewed as functions from inputs to outputs. The parallel between human beings and digital machines need not hold in every respect for such an analogy to be acceptable, however, since inanimate machines and human beings are otherwise very different kinds of things in innumerable respects.

As a consequence, there is a tendency to focus upon those processes that appear to be most likely to be unaffected by the differences between animate and inanimate things, where the subset of processes on which the most attention has focused are those that involve the processing of data, information, and knowledge. Indeed, the processes of logical and mathematical reasoning ("deductive inference"), of natural and technical language utilization ("linguistic capability"), and of visual and sensory information acquisition ("perceptual inference") have tended to receive considerable attention. Digital machines may not ingest, digest, or otherwise process food, water, and spirits as humans do, but these (presumably non-computational) processes can be set aside in favor of those "cognitive" processes they happen to share.

This more restricted focus suggests that AI is not so much attempting to create an artificial "human being" as an artificial "thinking thing", a prospect that theologians may find less menacing. In particular, Charniak and McDermott [(1985), p. 7] advance a sketch of what they think AI is really all about:

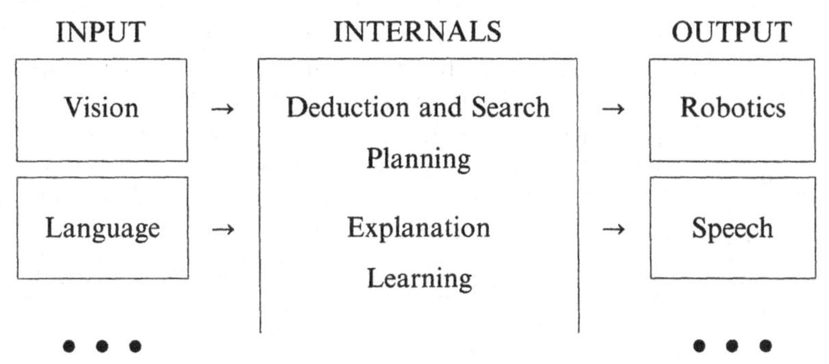

Fig. 2. Mental Faculties.

Here, "INTERNALS" stands for processes going on inside a human being (or inside a digital machine) and the sets of dots "..." and the open box imply that there may be more going on inside and outside the black box than has been portrayed. But bear in mind that, since digital machines are what AI researchers have to work with, any cognitive process that is not a computational process cannot be a computer process, but it may exist, nevertheless.

A great deal of excitement has been created by the claim that all cognitive processes are computational [see, for example, Pylyshyn (1984)]. This contention, of course, might be true or it might be false – a question whose answer we are going to discover. What is already clear, however, is that a description of AI as "the study of mental faculties through the use of computational models" harbors an important equivocation, since these "models" themselves might be of one or another kind, namely: there are models that *simulate* (by effecting the right functions from inputs to outputs) and models that *replicate* (by effecting the right functions by means of the very same – or similar – processes). Simply put, this means that machines might be able to do some of the things that humans can do (add, subtract, etc.), but they may or may not be doing those things the same way that humans do them.

THE AIM OF THIS INQUIRY

Models that simulate and models that replicate, as they are intended to be understood here, achieve those effects independently of consideration for the material (or "medium") by means of which they are attained. Another category of models, therefore, consists of those that *emulate* (by effecting the right functions by means of the same – or similar – processes implemented within the same medium). A relation of emulation between systems of different kinds thus entails that they be constituted of similar material (or components), which, in the case of humans, would involve flesh, blood, and nerves. Indeed, since digital machines are made of electronic components while human beings (leaving bionic men and women to one side) are not, it would be an inappropriate imposition to insist upon emulation as AI's goal.

Such a requirement, after all, would have the consequence of begging the question with regard to the fundamental issue of whether machines can have minds or not, since it would imply that only human beings are capable of emulating the cognitive processes of other human beings. Almost no one, however, challenges the capacity for AI to achieve the objective of simulating the cognitive processes of human beings across a broad range of species of reasoning, including, for example, number crunching. It would be odd, indeed, to doubt whether digital machines can add, subtract, etc., in the mode of simulation, since even hand-held calculators can generate the right output from the right input. The crucial question concerns whether machines could ever be constructed that would add, subtract, etc., in the mode of replication.

There are those, such as Fred Dretske (1985), who go so far as to deny that computers can add, subtract, etc., after all. But in adopting his position, Dretske is not suggesting that entering the sequence "2", "+", "2" with a handheld calculator could not produce "4" as its result. On the contrary, Dretske would concede that computers can simulate human reasoning while denying that that they can replicate human reasoning processes. The difficulty – and it is, of course, rather grave – becomes that of establishing a suitable evidential warrant for arriving at a conclusion concerning whether digital machines might or might not be able to replicate the cognitive processes of human beings. This question, moreover, is not one of technology, since there are few who would want to claim that current machines are capable of realizing such a goal today. Properly conceived, the question is one of physical possibility.

From this perspective, the purpose of this inquiry is to assess the field of artificial intelligence with respect to its scope and limits. Thus, one of the most pressing issues that we shall confront is whether or not machines have the potential for replicating human cognitive processes. Since this question can only be answered if we have some idea of the nature of human cognitive processes themselves, the character of mentality itself has to be addressed. Thus, the next two chapters are devoted to investigating the nature of mentality with respect to humans, other animals, and machines. The pursuit of this issue leads to an elaboration of the conception of minds as semiotic (or "sign-using") systems and to a comparison of this conception with the physical symbol system conception advanced by Alan Newell and Herbert Simon.

The arguments presented in Chapter 2 indicate that digital machines can qualify as symbol systems in Newell and Simon's sense, but that symbol systems of this kind are not semiotic systems. Even though these machines are capable of symbol processing in Newell and Simon's sense, they do not therefore qualify as the possessors of mentality. These issues are carried further in Chapter 3, moreover, where computational, representational, and dispositional conceptions are explored. Additional reasons are advanced for doubting that computers as currently conceived could possibly satisfy the desiderata required of thinking things. The possibility that machines of a different design (implementing parallel processing by means of connectionist architecture, for example) might have minds, however, cannot be entirely ruled out.

Chapter 2 emphasizes the nature of mentality, while Chapter 3 concerns the nature of language. The position developed in these chapters maintains that mentality involves a triadic relation between signs, what they stand for, and sign users; that computational conceptions reduce language and mentality to the manipulation of possibly meaningless marks; that even representational conceptions cannot salvage the standard conception, because they cannot account for the meaning of the primitive elements of language; and that only a pragmatical conception – that emphasizes the interrelations between signs, what they stand for, and sign users – can provide an adequate resolution of these difficulties. Thus, the symbol system conception may be suitable for digital machines but does not reflect the mentality of human beings.

Fortunately, the potential benefits of AI are not restricted to those that would be available if digital machines could replicate the mental processes of human beings. The representation of knowledge, the development of expert systems, the simulation and modeling of natural and of behavioral phenomena, and the rapid processing of vast quantities of data, information, and knowledge are destined to assume a major role in shaping the future of our species. In order to comprehend the full range of AI capabilities, therefore, it is indispensable to acquire an understanding of the nature of knowledge. Chapter 4 provides an introduction to the theory of knowledge and aims at the compact presentation of the crucial distinctions essential to this domain.

Perhaps the most important distinction that arises here, moreover, concerns the difference between natural and artificial language; for the results of the preceding chapters, which undermine the prospects for digital machines to understand natural language, reinforce their potential for securing enormous benefits through the utilization of artificial language. The level of discussion found in this chapter may be too detailed for beginners and not detailed enough for experts, but it affords a point of departure for an analysis of "knowledge" within the context of AI as presented in Chapter 5. The differences between ordinary and scientific knowledge are explored, including the nature of common sense and of defeasible reasoning. The implications of this approach for the frame problem are given initial consideration.

Chapter 6 focuses upon the nature of expert systems, with special concern for epistemic aspects of their construction and utilization. Even though expert systems are not in fashion with AI researchers preoccupied with the character of cognitive processes, they afford an excellent illustration of the extent to which AI can produce valuable products within the scope of simulation. Indeed, I think that expert systems will prove to be one of the most enduring legacies of the AI revolution. Chapter 7 continues this exploration by applying the distinctions introduced in previous chapters to understanding the distinctive characteristics of some of the most widely used modes of knowledge representation – semantic networks, predicate calculi, and scripts and frames – with special concern for their varied strengths and weaknesses.

Chapter 8 changes the cadence by taking a look at the epistemic foundations of programs themselves by exploring distinctions between validity and soundness, algorithms and programs, abstract entities and physical systems, and pure and applied mathematics. Some readers may want to bypass this chapter, but others will appreciate its general significance for understanding the epistemic limitations of computers and their programs. Chapter 9, finally, returns to the problems of cognition with which the book began, turning once more to the differences between humans and machines concerning the nature of mentality. This summation of the situation is then extended to an exploration of the relationship between bodies and minds and the nature of our knowledge of other minds, thus completing an investigation of the three great problems of the philosophy of mind as they arise within the context of AI.

The conception that this work is intended to convey, therefore, is that the scope of AI should not be limited by concern for the problem of replication. Even if digital machines are restricted to the utilization of artificial languages that may be significant for the users of those systems but not for use by those systems themselves, the contributions that AI products can extend to the enhancement of human existence are very important, indeed. Not the least of the benefits that this book ought to provide, I should add, is a more systematic appreciation for the influence which presuppositions – especially with respect to methodology – can exert upon our theoretical investigations. Thus, it ought to be emphasized that the conception of mentality advocated here is strongly non-behavioristic, non-extensional, and non-reductionistic, where these notions are to be understood as the following passages explain.

Behaviorism. Various conceptions concerning the character of scientific language have pivoted about the empirical testability or the observational accessibility of hypotheses and theories. A distinction between three kinds of non-logical terms has often been drawn, where predicates that designate (or "stand for") properties are classified on the basis of whether the properties they stand for are observational, dispositional, or theoretical in kind. Richard Rudner (1966), for example, has proposed this classification scheme:

(D1) observational predicates =df predicates that stand for observable properties of observable entities;

(D2) dispositional predicates =df predicates that stand for unobservable properties of observable entities; and,

(D3) theoretical predicates =df predicates that stand for unobservable properties of unobservable entities;

where every predicate should fall into one and only one of these categories.

But a long-standing debate has raged over the merits of predicates making reference to unobservable properties within the context of empirical inquiries. Practically all parties to this dispute tend to acknowledge the fundamental function fulfilled by observational language, which is used to describe the contents of (more or less) direct experience. The problem thus arises over the status of unobservable properties, where much of the war has been waged over the scientific standing and scientific significance of dispositional predicates, which are intended to ascribe to things habits or tendencies to display specific kinds of behavior under specific conditions. The participants in this debate have included Gilbert Ryle in philosophy and B. F. Skinner in psychology, but its origin traces back to David Hume. [See, for example, Carnap (1936-37), Ryle (1949), and Skinner (1953).]

Since dispositions are tendencies to display specific kinds of behavior under specific conditions, it looks as though it might be possible to settle the debate over the status of dispositional language, at least, by establishing that these unobservable properties are explicitly definable by means of observational predicates and truth-functional logical connectives, such as the "if ___ then . . ." conditional understood as the *material* conditional. The predicate "x is hungry" can then be defined in the following fashion:

(D4) x is hungry at time t =df if x is given food at time t, then x eats that food by time t^*;

where t^* equals t plus some suitable (specific) interval, permitting time for that behavior to be displayed. The promise of this approach in offering a satisfactory solution to these problems of meaning and significance was so great that it became a cornerstone for the position that is known as *behaviorism,* which endeavors to eliminate appeals to unobservables by strict adherence to this methodology [see, for example, Block (1980)].

The imposition of these rigid epistemic requirements upon scientific definitions has proven highly problematical, not least of all because the material conditional, although well-defined, does not possess the logical properties appropriate to dispositional predicates. In particular, since a material conditional is true when either its antecedent is false or its consequent is true, when "x is hungry" is defined by (D4), its definiens will be satisfied by anything x at any time t that it is not being given food as well as by anything that is given food at that time and eats that food during the specified interval. If a leather chair or a delicate antique does not happen to be given food at a certain time t, then it turns out to be hungry at that time, as a consequence of the form adopted for proper definitions.

The problem for leather chairs and delicate antiques, however, can be disposed of by restricting the class of kinds of things that might or might not be hungry to animate things, so long as their possession of this property could be ascertained on the basis of experiential grounds alone. For the underlying rationale for behaviorism's emphasis on *stimuli* and *responses* arises from their construction as publicly observable antecedents and as publicly observable consequents. But the problem turned out to be even-more difficult than that, since there appear to be other factors besides its availability that influence whether or not a human being, for example, is disposed to eat when presented with food. If a person were participating in a religious fast, scrupulously adhering to a rigid diet, or otherwise morally inhibited from consuming food, he might very well not eat food when it was presented, even though he was indeed hungry. To cope with problems of this kind, however, it would be necessary to assume the existence of other unobservable properties of x, which tends to defeat the program.

Extensionality. There were skillful efforts to overcome these difficulties, some of the most ingenious of which were undertaken by Rudolf Carnap in (1936-37). Yet even Carnap's ingenuity was not enough to salvage the behaviorist platform, for at least two different reasons. The first has already been implied above, namely: that the causal factors that tend to influence human behavior cannot be reduced to those that are accessible to observation in the form of publicly observable stimuli and publicly observable responses. If we tend to accept ordinary "folk" psychology, a more adequate inventory of the inner states of human beings would include not only motives and beliefs but also ethics, abilities, capabilities, and opportunities. But the complex causal interaction of factors of these kinds renders their presence or absence subject to indirect modes of inference rather than to direct observation. The epistemic resources that behaviorism embraced were inadequate to cope with the complexities of the subject of behavior.

Even more striking than the limitations that behaviorism imposed on the non-logical terms that it wanted to employ were those discovered to attend its logical resources, in particular. Coping with what would happen if x were to be given food (whether or not x ever is) and with what would have happened had x been given food (when x had not) requires the use of *subjunctive* conditionals and *counterfactual* conditionals – where the latter are subjunctives with false antecedents – in intensional logic. Even Carnap eventually arrived at the conclusion that the definition of dispositional predicates could only be accomplished by reaching beyond the resources of material conditionals and of extensional logic. Thus, the behaviorist program for the elimination of theoretical and dispositional language could not be sustained. [For an illuminating survey, see Hempel (1965), pp. 101-122.]

In its full dimensions, therefore, the behaviorist position encountered three distinct problems. These can be formalized by utilizing '____ $\rightarrow$ . . .' as the material-conditional sign and '____ $\Rightarrow$ . . .' as the subjunctive-conditional sign. Then let "Sxt" stand for x's exposure to stimulus S at time t, "Rxt^*" stand for x's display of response R at time t^*, and "$F1$, $F2$, . . . , Fm" stand for other properties (which are not necessarily directly accessible). For defining dispositional predicates, (1) "$Sxt \rightarrow Rxt^*$" is logically flawed because of the limitations of the material conditional; (2) "$Sxt \Rightarrow Rxt^*$" is empirically flawed because of the restriction to observable causal factors; yet, (3) "$(Sxt \;\&\; F1xt \;\&\; \ldots \;\&\; Fmxt) \Rightarrow Rxt^*$", which at least appears to be headed in the right direction, is not compatible with the behaviorist program. Even the subjunctive turns out to not be strong enough to fulfill its intended role and ultimately requires replacement by a causal conditional. [For more discussion of this and related issues, see Fetzer (1978), (1985a).]

What often tends to be overlooked about the material conditional is that its use presumes no special kind of connection between its antecedent and its consequent. They do not have to be related by virtue of meaning, or of causality, or of lawfulness, or of any consideration other than merely the *truth values* of those sentences alone. Hence, whenever the antecedent of a material conditional happens to be false, that entire conditional will be true no matter how unrelated or disconnected the subjects of its antecedent and its consequent may be: "If nine were even, Reagan would be President" and "If nine were even, Reagan would not be President" are both trivially true if interpreted as material conditionals, because both their antecedents are false!

While logicians have long since grown accustomed to this result, many of those laboring in other fields find this mystifying, with suitable justification. Ordinary language, educated intuition, and common sense are violated by a failure to distinguish between conditionals with false antecedents that are true and conditionals with false antecedents that are false. But extensional logic does not permit them to be distinguished, since *all* material conditionals with false antecedents are true. The need for stronger modes of conditionality should therefore be apparent, where the resources provided by *intensional* (or "non-truth-functional") logic far exceed those of *extensional* (or "truth-functional") logic, where the use of a subjunctive conditional, for example, entails but is not entailed by a corresponding material conditional. Indeed, the only case for which the truth value of a subjunctive is determined by the truth value of the correponding material conditional is when that material conditional happens to be false [cf. Fetzer and Nute (1979), (1980)].

The use of material conditionals, moreover, implies that an outcome response will occur whenever the input stimulus occurs, which will not happen when the relationship between them is one of probability instead. A behavioristic approach that accommodates probabilistic relationships by invoking the notion of "probability of response", therefore, has been elaborated in the work of B. F. Skinner [(1953), (1972)]. From Skinner's point of view, a disposition like hunger should be characterized, not as an invariable tendency to eat whenever food is presented, but as a high probability:

(D5) x is hungry at time t =df the probability that x will eat food by time t^* when given food at t is high.

Since "probability" must be interpreted as an extensional relative frequency rather than as an intensional causal propensity in order to preserve its "scientific respectability", however, this account fall prey to the same problems that undermine material conditionals, a point to which we return in Part II.

Reductionism. These distinctions, as I have hinted above, are meant to provide an introduction to some issues that lie just beneath the surface of the problems that are the principal subjects of our discussion in the following. For I shall argue that the wrong conclusions have sometimes been embraced because of mistaken commitments to untenable methodologies, especially those of behaviorism and of extensionality. The point, therefore, is to provide enough background to appreciate the reasons why these methodologies are open to dispute without attempting to exhaust the problems that they raise, while suggesting resources for further exploration. Thus, the last of these that I want to mention here is also the most general, for both of the others can be viewed as special cases of the program known as *reductionism*.

My reason for saying so is that, when behaviorism as it has been defined functions as a standard of acceptability for research in psychology, it serves as an attempt to reduce the unobservable to the observable within the field of psychology. And when extensionality as it has been defined functions as a standard of acceptability for scientific language, it serves as an attempt to reduce the intensional to the extensional within all fields of science. For, in a similar fashion, reductionism can be viewed as any effort to reduce what is complex to what is simple, largely driven by the spirit of Occam's Razor, a methodological maxim asserting that entities should not be multiplied beyond necessity: "It is in vain to do by many what could be done by fewer!" [On Occam's Razor and its significance, cf. Smart (1984) and Fetzer (1984).]

Occam's Razor thus suggests that we ought to prefer simpler over more complex theories, which provides leverage in moving toward less complex theories. But reductionism, thus understood, suffers from one critical flaw. For surely we must prefer simpler to complex theories only when they are adequate. Behaviorism, after all, provides a simpler account of the nature of psychological properties than its theoretical alternatives, but it remains hopelessly inadequate, nevertheless. Extensionality, likewise, provides a simpler account of the nature of scientific language than its theoretical alternatives, but it remains hopelessly inadequate, nevertheless. Reductionism is an appealing conception, but it is not necessarily always justifiable.

If we want to define dispositional predicates like "hunger", which are properties of observable entities such as human beings, the appropriate place to begin would be with material conditionals. Once we have discovered the consequence that people turn out to be hungry merely because they are not being subjected to the appropriate test, however, it becomes imperative to consider alternative constructions. And if there simply is no simple relationship between stimulus and response for human beings because they are complex causal systems whose behavior is affected by inner states that are not amenable to direct observation, then we ought to embrace a more sophisticated methodology that is equal to the task that we would impose upon it. And if there should be no simple way to relate the mental to the physical, that is something that we shall have to accept.

Perhaps a more appropriate conception of Occam's advice would be to construe his concern as a drive for "elegance", in the sense of striving to derive maximal sets of consequences from minimal sets of assumptions in order to optimize, say, explanatory power in science or theoretical significance in philosophy. For, if this were the case, then simpler theories would only be preferred theories when they provide an acceptable account of the phenomena with which they are intended to deal. When matters are complex, Occam's Razor would no longer have us embrace simple theories, but only those that are as simple as they can be while remaining adequate to cope with the demands of a complex world. The problems we confront in dealing with the scope and limits of AI are complex. Perhaps in place of simple theories we ought to be looking for elegant theories instead.

2. SYMBOL SYSTEMS AND SEMIOTIC SYSTEMS

Since much of the debate about the mentality of machines depends upon or arises from analogical reasoning, it might be worthwhile to review what this species of reasoning is all about. Generally speaking, reasoning by analogy occurs when two things (or kinds of things) are compared and the inference is drawn that, since one – the more familiar case – has some particular property, the other – less familiar case – does too. Reasoning by analogy can be faulty (or "fallacious") when (i) there are more relevant differences than there are relevant similarities drawing these cases together, (ii) the inference is taken to be completely conclusive, when at most it can be strongly supportive, or (iii) although there are more relevant similarities than there are relevant differences, there is some important (crucial or critical) property distinguishing between them, an instance of what may be described as the *Principle of Minimality* (to which we shall later have the occasion to return).

Among the principal difficulties in employing analogical reasoning, however, is figuring out how to ascertain which properties are the relevant ones and which are not. If the properties that the familiar case is known to possess are its *reference properties* and the property of interest is an *attribute*, then every reference property whose presence or absence "makes a differerence" to the presence or absence of the attribute is an evidentially relevant property. Unfortunately, there are two quite different standards for measuring whether or not a difference is made by the presence or absence of a reference property, namely: the *frequency* and the *propensity* criteria [cf. Fetzer (1981), Part II]. The frequency criterion is purely descriptive in rendering a property statistically relevant to an attribute whenever the frequency with which it occurs varies with that specific property's occurrence. The propensity criterion, by comparison, is highly theoretical in rendering a property causally relevant to an attribute whenever the causal tendency for it to occur varies with that specific property's presence. Their differences will be the subject of futher exploration when we turn to Chapters 5 and 6.

These considerations are important to comprehend, precisely because so many of the arguments we are considering implicitly involve or tacitly presuppose analogical reasoning comparing human beings and digital machines. The foundations of these analogies require critical scrutiny, especially in the case of the symbol-system conception offered by Allen Newell and Herbert Simon and in the case of the computational conception of language and mentality. They may – without exaggeration – be described as the "twin pillars" of contemporary research in AI. Newell and Simon's position involves reasoning from the features of digital machines to those of human beings. The computational conception extends its analysis of the internal processes of these machines to those of human beings. In order to display the limitations of these positions, therefore, in this chapter and the next, I attempt to establish that the analogies upon which they rest are faulty ones. In doing so, I shall introduce a more adequate theory of the nature of mind.

According to the *physical symbol system* conception advanced by Newell and Simon (1976), physical symbol systems are machines – possibly human – that process symbol structures across time. For Newell and Simon, AI deals with the development and construction of special kinds of physical systems that employ symbols to represent and to utilize information or knowledge, a position that is often either explicitly endorsed or tacitly adopted by authors and scholars at work within this field [such as Nii et al. (1982) and Buchanan (1985)]. Indeed, this perspective has been said to be "the heart of research in artificial intelligence" [Rich (1983), p. 3], a view that appears to be representative of the standing of their position within the AI community at large. It is therefore appropriate to consider Newell and Simon's account in detail.

The tenability of this conception, of course, obviously depends upon the notions of system, of physical system, and of symbol system that it reflects. Newell and Simon, for example, tend to assume that symbols may be used to designate any expression whatever, where these expressions can be created and modified in arbitrary ways. They further expand upon their conception by proposing the "physical symbol system hypothesis" – the conjecture that physical symbol systems satisfy the necessary and sufficient conditions for "intelligent action". Indeed, it is the physical symbol system *hypothesis* as opposed to the physical symbol system *conception* that represents Newell and Simon's inference from the properties of these machines to the properties of human beings. If their analogy is a faulty one, therefore, it must be by virtue of the differences and the similarities distinguishing these cases.

Not the least important question to consider, however, is what an ideal framework of this kind ought to be able to provide. In particular, as their physical symbol system hypothesis itself suggests, the conception of symbol systems is intended to clarify the relationship between symbol processing and deliberate behavior, in some very broad sense. Indeed, it would seem to be a reasonable expectation that an ideal framework of this kind ought to have the capacity to shed light on the general character of the causal connections that obtain between mental activity and behavioral tendencies to whatever extent they occur as causes or as effects of the production and the utilization of symbols or of their counterparts within an alternative conception. For, as Jerry Fodor has observed, "(A) cognitive theory seeks to connect the *intensional* properties of mental states with their *causal* properties vis-a-vis behavior" [Fodor (1980), p. 325]. Indeed, any account that fails to satisfy this condition, whatever its other merits, would not be a "theory of mind".

The argument developed here is intended to undermine the adequacy of Newell and Simon's conception, not in relation to digital machines, but for understanding mentality as it occurs in human beings. I will examine these issues from the point of view of a theory of mind based upon the theory of signs proposed by Charles S. Peirce (1839-1914). According to the account that I shall elaborate, minds are defined as *semiotic systems*, among which systems that use symbols (which are not the same thing as symbol systems in Newell and Simon's sense) are but one among three basic kinds. As semiotic systems, minds are sign-using systems that have the capacity to create or to utilize signs, where this capability might be either naturally produced or artificially contrived. The result is a conception of mental activity that promises to clarify and illuminate the similarities and differences between semiotic systems of various kinds, no matter whether human, other animal, or machine – and thereby shed substantial light upon the foundations of AI.

PEIRCE'S THEORY OF SIGNS

It is not uncommon to suppose that there is a fundamental relationship between language and thought, as when it is assumed that all thinking takes place in language. But it seems to me that there is a deeper view that cuts into this problem in a way in which the conception of an intimate connection between language and thought cannot. This is a theory of the nature of mind that arises from reflection upon the theory of signs (or "semiotic theory") advanced by Peirce. The most fundamental concept Peirce advanced is that of a *sign* as a something that stands for something (else) in some respect or other for somebody. Signhood thus involves a three-place (triadic) relation:

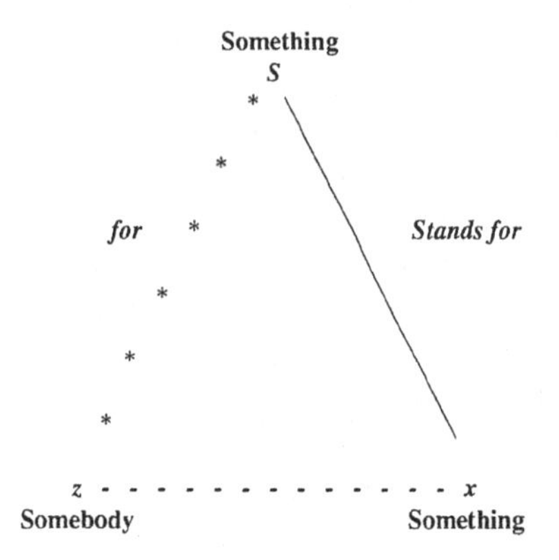

Fig. 3. The Sign Relation I.

Peirce further distinguished between three principal areas of semiotic inquiry: the study of the relations signs bear to other signs; the study of the relations signs bear to that for which they stand; and the study of the relations obtaining between signs, what they stand for, and sign users. While Peirce himself referred to these dimensions of semiotic as "pure grammar", "logic proper" and "pure rhetoric", they are more familiar under the designations of "syntax", "semantics", and "pragmatics", respectively, terms first introduced by Charles Morris (1938) but which have since become standard.

Within the domain of semantics, Peirce identified three ways in which a sign might stand for that for which it stands, thereby generating a classification of three kinds of signs. Any things that stand for that for which they stand by virtue of a relation of resemblance between those signs themselves and that for which they stand are known as "icons". Statues, portraits, and photographs are icons in this sense, when they create in the mind of a sign user another – equivalent or more developed – sign that stands in the same relation to that for which they stand as do the original signs creating them. Any things that stand for that for which they stand by virtue of being either causes or effects of that for which they stand are known as "indices". Dark clouds that suggest rain, red spots that indicate measles, and ashes that remain from a fire are typical indices in this sense. Things that stand for that for which they stand by virtue of conventional agreements or by virtue of habitual associations between those signs and that for which they stand are known as "symbols". Most of the words that occur in ordinary language, such as "chair" and "horse" – which neither

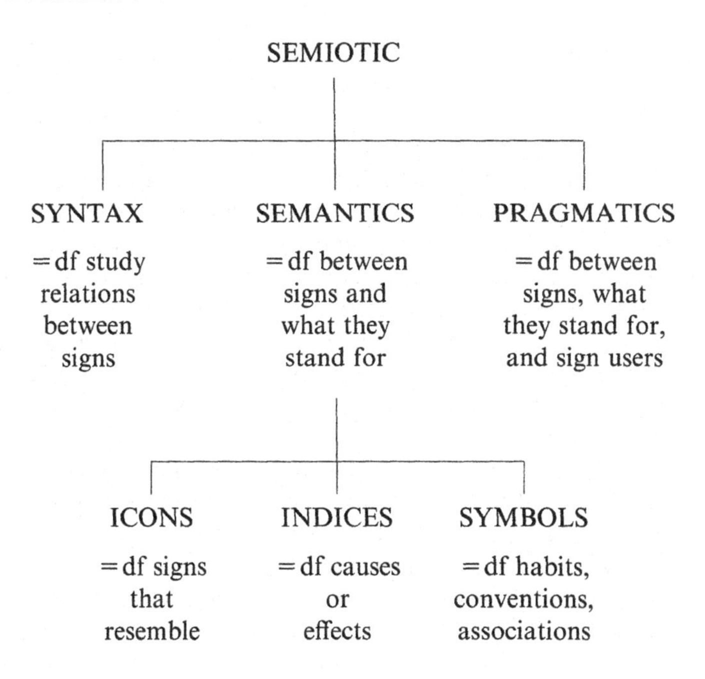

Fig. 4. Peirce's Theory of Signs.

resemble nor are either causes or effects of that for which they stand – are symbols in this technical sense. [See, for example, Hartshorne and Weiss (1960), 2.247-249 and 2.274-308.]

There is great utility in the employment of symbols by the members of a sign-using community, of course, since, purely as a matter of conventional agreement, almost anything could be used to stand for almost anything else under the specific conditions that might be established by the practices that would govern the use of those signs within a community of sign users. The kinds of ways in which icons and indices can stand for that for which they stand may be thought of as *natural* modes, because relations of resemblance and of cause-and-effect are there in nature, whether we notice them or not, whereas conventional or habitual associations or agreements are only there if we create them. Nevertheless, it would be a mistake to make too much of this distinction, because things might be alike or unalike in infinitely many ways, where two things qualify as of a common kind if they share common properties from a certain point of view, which may or may not be easy to ascertain.

The most important conception underlying this reflection is that of the difference between *types* and *tokens*, where "types" consist of specifications of kinds of things, while "tokens" occur as their instances. The color blue, for example, when appropriately specified (by means of color charts, in terms of angstrom units, or otherwise), can have any number of instances, including bowling balls and tennis shoes, where each instance of that color is a token of that type. Similar considerations obtain for sizes, shapes, weights, and all the rest: any property (or pattern) that can have distinct instances may be characterized as a type (or kind), where the instances of that type (or kind) qualify as its tokens. The necessary and sufficient conditions for a token to qualify as a token of a certain type, moreover, are typically referred to as the *intension* (or "meaning") of that type, while the class of all things that satisfy those conditions is typically referred to as its *extension* (or "reference"). This distinction applies to icons, to indices, and to symbols alike, where identifying things as tokens of a type presupposes a point of view, which might not be the same thing as adopting some semiotic framework.

The importance of perspective can be exemplified. My feather pillow and your iron frying-pan both weigh less than 7 tons, are not located on top of Mt. Everest, and do not look like Lyndon Johnson. Paintings by Rubens, Modigliani, and Picasso clearly tend to suggest that relations of resemblance presume a point of view. A plastic model of the battleship *Missouri* may be like the "Big Mo" with respect to the relative placement of its turrets and bulwarks, yet fail to reflect other properties – such as the mobility and firepower – of the real thing. Whether a specific set of similarities qualifies as resemblance thus depends upon and varies with the adoption of some perspective. And, indeed, even events are described as "causes" and as "effects" in relation to implicit commitments to natural laws and to scientific theories. Relations of resemblance and causation have to be recognized to be utilized, where the specification of a complex type can be a very complex procedure.

In thinking about the nature of mind, it seems plausible that Peirce's theory of signs might provide us with clues. In particular, reflecting upon the conception of a sign as a something that stands for something (else) in some respect or other for somebody, it appears to be a presumption to assume that those somethings for which something can stand for something (else) in some respect or other must be "somebodies". It would be better to suppose, in a non-question-begging way, that these are "somethings", without taking for granted that the kind of thing that these somethings are has to be human. In reasoning about the kinds of things that these somethings might be, the possibility suggests itself that these things for which things can stand for other things should themselves be viewed as "minds". [Morris (1938), p. 1, intimates that (human) mentality and the use of signs may be closely connected.] The conception that I shall adopt, therefore, is that *minds* are things that are capable of utilizing signs (or are "sign users"), where semiotic systems in this sense are causal systems of a special kind.

Abstract and Physical Systems. In defense of this position, I need to explain, first, what it means to be a system, what it means to be a causal system, and what it means to be the special kind of causal system that is semiotic, and, second, what, if anything, makes this approach more appropriate than the Newell and Simon conception, especially with respect to the foundations of artificial intelligence. The arguments that follow suggest that Newell and Simon's account harbors an important equivocation, insofar as it fails to define the difference between a set of symbols that is significant for a *user of* a machine, in which case there is a semiotic relationship between the symbols, what they stand for, and the symbol user (where the user is not identical with the machine), and a set of symbols that is significant for *use by* a machine, in which case there is a semiotic relationship between the symbols, what they stand for, and the symbol user (where the user is identical with the machine). The critical difference between symbol and semiotic systems emerges at this point. Let us begin with the nature of systems in general.

A *system* may consist of a collection of things – numbers and operators, sticks and stones, or whatever – that instantiates a fixed arrangement. This means that a set of parts becomes a system of a certain kind by instantiating a set of specific relations – logical, causal, or whatever – between those parts. A system may be functional or non-functional with respect to various inputs and outputs, where the difference depends upon whether or not that system responds to those inputs or produces those outputs. Whether or not this is the case does not always result from intentions or by design. In any case, for a fixed input – such as assigning seven as the value of some variable, resting a twelve-pound weight on its top, and so on – a specific system will tend to produce a specific output (which need not be unique to its output class) – such as yielding fourteen as an answer, collapsing in a heap, and so forth. Differences in the kinds of outputs that a system will produce under the influence of inputs of the same kind, therefore, can serve to distinguish between systems of different kinds. Since systems that differ in their construction could produce the same kinds of outputs and probabilistic systems can produce more than one kind of output under the influence of the same inputs, differences in outputs, in the former case, or in output classes, in the latter, are sufficient but not necessary for systems to differ.

The "functional/non-functional" distinction is important to appreciate. For most arrangements of lighting and electricity, for example, whistling in the dark does not make the lights come on, because of which all those systems are non-functional with respect to that input. It is not difficult to imagine an ingenious inventor, however, who could create a system of a different kind for which that system produces that output if subject to that input. It is tempting to make the assumption that systems that are functional with respect to the same classes of inputs and the same classes of outputs – leaving aside, for the moment, differences that arise from probabilities – should be viewed as systems of the same kind *K*. Whether or not this is a good idea, however, tends to depend upon one's point of view, an issue that we are going to pursue. Notice, in particular, that such an assumption only succeeds so long as whatever is going on in between inputs and outputs does not matter.

For a system to be a *causal* system means that it is a system of things in space/time between which causal relations can obtain. An *abstract* system, by comparison, is a system of things not in space/time between which only logical relations can obtain. This conception of a causal system thus bears a strong affinity to Newell and Simon's conception of a *physical* system, which is a system governed by the laws of nature. Indeed, to the extent to which the laws of nature include probabilistic as well as deterministic laws, both of these conceptions should be interpreted broadly (but not reductionistically, because the non-occurrence of "emergent" properties – possibly including at least some mental phenomena – is not a feature of the interpretation that is intended). Since systems of neither kind are restricted to inanimate as opposed to animate systems, the advantage of employing "causal system" in lieu of "physical system" is that the former explicitly permits the possibility of systems of a type whose existence is only implicitly conceded by the latter.

Within the class of causal (or of physical) systems, therefore, two subclasses require differentiation. Causal systems whose relevant properties are only incompletely specified may be referred to as "open", while systems whose relevant properties are completely specified are regarded as "closed". Then distinctions may be drawn between two kinds of closed causal systems as suggested above, namely: those for which, given the same input, the same output invariably occurs (without exception), and those for which, given the same input, one or another output within the same class of outputs invariably occurs (without exception). Systems of the first kind, accordingly, may be characterized as *deterministic* causal systems, while those of the second kind are *indeterministic* (or "probabilistic") causal systems [Fetzer (1981), Part I].

This last distinction is not the same as another drawn in computational theory between between deterministic and non-deterministic finite automata, such as parsing schemata, which represent paths from grammatical rules (normally called "productions") to sentences (often called "terminal strings"), for which more than one production sequence (parse tree) is possible, where *human choice* influences the path (or parse) selected [Cohen (1986), pp. 142-145]. While Newell and Simon acknowledge this distinction, the systems to which it applies, strictly speaking, are special kinds of "open" rather than of "closed" causal systems, precisely because the complete sets of factors that influence the outcomes of production sequences are not explicitly specified.

The conception of abstract systems, no doubt, also merits additional discussion, where purely formal systems – the systems of the natural numbers, of the real numbers, and the like – are presumptive examples thereof. While abstract systems, unlike physical systems, are not in space/time and cannot exercise any causal influence upon the course of events during the world's history, this result does not imply that, say, inscriptions of numerals – as their representatives within space/time – cannot exercise causal influence as well. Indeed, since chalkmarks on blackboards affect the production of pencilmarks in notebooks, where some of these chalkmarks happen to be numerals, such a thesis would be difficult to defend. We are able to reason about abstract systems by means of physical signs that can represent them.

For a causal system to be a semiotic system, of course, it must be a system for which something can stand for something (else) in some respect or other, where such a something (sign) can affect the (actual or potential) behavior of that system. In order to allow for the occurrence of dreams, of daydreams, and of other mental states as potential outcomes (responses or effects) of possible inputs (stimuli or trials) – or as potential causes (inputs or stimuli) of possible outcomes (responses or effects) – behavior itself requires a remarkably broad and encompassing interpretation. A conception that accommodates this possibility is that of *behavior* as any internal or external effect of any internal or external cause. This account circumvents the arbitrary exclusion of internal (or private) occurrences from the class of possible responses, which are often restricted to external (or public) happenings. Indeed, from this point of view, it should be apparent that for something to affect the behavior of a causal system does not mean that it has to be a sign for that system, which poses the major problem for the semiotic approach: distinguishing semiotic causal systems from other kinds of causal systems.

Causal and Semiotic Systems. To appreciate the dimensions of this difficulty, consider that if the capacity for the (actual or potential) behavior of a system to be affected by something were enough for that system to qualify as semiotic, the class of semiotic systems would be coextensive with the class of causal systems: they then would have all and only the same members. If even one member of the class of causal systems does not qualify as a member of the class of semiotic systems, however, then this identification cannot be sustained. Since my coffee cup, your reading glasses, and numerous other things – including sticks and stones – are systems whose (actual and potential) behavior can be influenced by innumerable causal factors, yet surely should not qualify as semiotic systems, something more had best be involved here. That something can affect the behavior of a causal system is not enough.

That a system's behavior can be affected by something is necessary, of course, but, in addition, the something must be functioning as a sign for that system. That such a sign stands for that for which it stands for that system must make a difference to the (actual or potential) behavior of that system, where this difference can be specified in terms of the various ways that such a system would behave, were such a sign to stand for something other than that for which it stands for that system (or, would have behaved, had such a sign stood for something other than that for that system). Were what a red light at an intersection stands for to change to what a green light at an intersection stands for (and conversely) for particular causal systems, including little old ladies but also fleeing felons, then that those signs now stand for things other than that for which they previously stood ought to have corresponding behavioral manifestations, which, in accordance with the definition for behavior we have adopted above, might be internal or external in kind.

Little old ladies who are not unable to see, for example, should now slow down and come to a complete stop at intersections when green lights appear, and release the brake and accelerate when red lights appear. Felons fleeing with the police in hot pursuit, by contrast, may still speed through, but they worry about it a bit more, which, within the present context, qualifies as a behavioral manifestation. Strictly speaking, changes in external behavior (with respect to outcome classes) are sufficient but not necessary, whereas changes in internal behavior (with respect to outcome classes) are necessary and sufficient for a sign to have changed its meaning for a system. Thus, a more exact formulation of the principle in question would state that, for any semiotic system, a sign S stands for something x for that system rather than for something else y if and only if the strength of the tendencies for the system to manifest behavior of some specific kind in the presence of S – no matter whether publicly displayed or not – differs from case to case, i.e., when S stands for x, those tendencies are not the very same as when S stands for y, a conception that we shall explore in greater detail in the following chapter.

This principle implies that a change that effects no change is no change at all in relation to the significance of a sign for a system. No change ought to occur when one token is exchanged for another token of the same type. Once again, however, considerations of perspective have to be factored in, since one dime (silver) need not stand for the same thing as another dime (silver and copper) for the same system when they are tokens of some of the same types, but not of others. Although this result appears agreeable enough, the principle of significance being proposed does not appear to be especially practical, since access to strengths of tendencies for behavior that may or may not be displayed is empirically testable in principle, but only indirectly measureable in practice [Fetzer (1981), (1986)]. This theoretical conception can be supplemented by an additional criterion for that purpose.

The measure being proposed is intended to provide an account of what it means for a sign to change its meaning (what it stands for) for a system, where the differences involved here may be subtle, minute, and all but imperceptible. For defining "sign", it would suffer from circularity in accounting for what it means for a sign to change its meaning while relying upon the concept of a sign itself. It does not provide a definition of what it is to be a sign or of what it is to be a semiotic system as such, but of what it is for a sign to change its meaning for a semiotic system. This difficulty, however, can be at least partially offset by appealing to what appears to be a suitable criterion for a system to be a semiotic system (for a thing to be a mind), namely: *the capacity to make a mistake*. For in order to make a mistake, something must take something to stand for something other than that for which it stands, a reliable evidential indicator that something has the capacity to take something to stand for something, which is the right result.

We should all find it reassuring to discover that the capacity to make a mistake – to mis-take something for other than that for which it stands – appears to afford conclusive evidence that something has a mind. That something must have the capacity to make a mistake, however, does not mean that it must actually make them as well. The concept of a divine mind that never makes mistakes – no matter whether as a matter of logical necessity or as a matter of lawful necessity for minds of that kind – after all, is not inconsistent [cf. Fetzer (1988a)]. The difference between mistakes and malfunctions, moreover, deserves to be emphasized, where mistakes are made by systems while remaining systems of the same kind, while malfunctions transform a system of one kind K into a system of another K^*. That a system makes a mistake is not meant to imply that its output classes, relative to its input classes, have been revised, but rather that, say, faulty reasoning has occurred, the false has been taken for the true, or something has been misclassified, which frequently occurs in perceptual and inductive inference.

Mistakes can occur using signs of any of the three basic kinds. In the case of icons, for example, the misidentification of the subject of a photograph, of a participant in a lineup, and the like, illustrate a few of the mistakes that can occur by virtue of taking a resemblance relation of one kind for a resemblance relation of another. In the case of indices, moreover, responding to the flash and flames on the set of a movie production as though it were a case of uncontrolled combustion would be an instance in which a "special effect" would be mistaken for a "normal effect" of its typical cause. In the case of symbols, indeed, things are decidedly more complicated insofar as the very existence of significant signs of this kind depends upon conventional agreements and habitual associations, which might, after all, be misunderstood because of ambiguity, because of synonymy, or because of context, matters that we shall pursue further in the chapter that follows.

THE VARIETIES OF SEMIOTIC SYSTEMS

The semiotic analysis of minds as semiotic systems invites the introduction of at least three different types (or kinds) of minds, where systems of Type I can utilize icons, systems of Type II can utilize icons and indices, and systems of Type III can utilize icons, indices, and symbols. If the conception of minds as semiotic systems is right-headed, at least in general, therefore, it would seem reasonable to conjecture that there are distinctive behavioral (psychological) criteria for systems of each of these different types. That is, if this approach is even approximately correct, then it should not be overly difficult to discover that links can be forged with psychological (behavioral) distinctions that relate to the categories thereby generated. In particular, there appear to be three kinds of learning (conditioning, whatever) distinctive to each of these three types of systems: semiotic systems of Type I exhibit type/token recognition, those of Type II exhibit conditioned association, and those of Type III exhibit instrumental conditioning, where behavior of these different kinds appears to be indicative that a system is of that type.

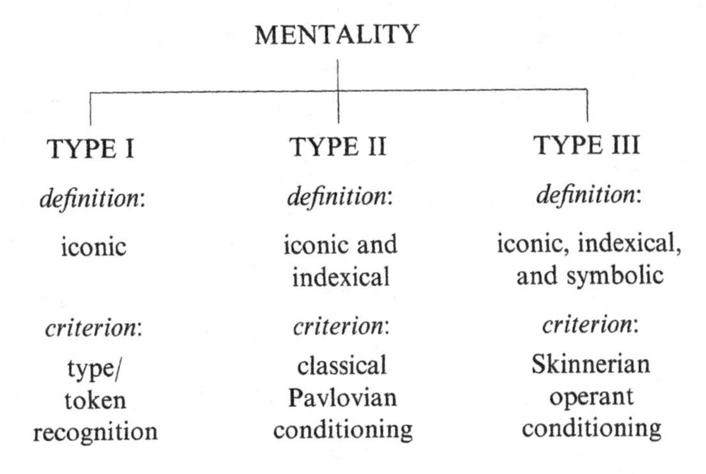

Fig. 5. Minds as Semiotic Systems.

Let us therefore begin by considering examples of semiotic systems of these three kinds and subsequently turn to the symbol system hypothesis itself.

Non-human animals provide useful examples of semiotic systems that display classical conditioning, for example, as systems of Type II that have the capacity to utilize indices as signs. Pavlov's experiments with dogs are illustrative here. Pavlov observed that dogs tend to salivate at the appearance of their food in the expectation of being fed and that, if a certain stimulus, such as a bell, were regularly sounded at the same time its food was brought in, a dog soon salivated at such a bell's sound whether its food was present or not. From a semiotic perspective, the food itself functions as a (relatively speaking) unconditioned stimulus for dogs, namely: as a sign that stands to the satiation of their hunger as a cause stands to its effects. Thus, when the sound of a bell functions as a conditioned stimulus for a dog, it similarly functions as a sign that stands to the satiation of its hunger as a cause stands to its effects. In the case of the unconditioned food stimulus, of course, the stimulus actually is a cause of hunger satiation, while in the case of the conditioned bell stimulus, it is not. This does not defeat the example, since it displays that dogs sometimes make mistakes.

Analogously, Skinner's familiar experiments with pigeons provide an apt illustration of semiotic systems of Type III that have the capacity to utilize symbols as signs. During the course of his research, Skinner found that pigeons kept in cages equipped with bars that would emit a pellet if pressed rapidly learned to depress the bar whenever they wanted some food. He also discovered that if, say, a system of lights was installed, such that a bar-press would now release a pellet only if the green light was on, they would soon refrain from pressing the bar, even when they were hungry, unless the green light was on. Once again, of course, the pigeon might have its expectations disappointed by pressing the bar when the apparatus has been changed (or the lab assistant forgot to set a switch, or whatever), which shows that pigeons are no smarter than dogs at avoiding mistakes.

Classical conditioning and operant conditioning, of course, are rather different kinds of learning. The connection between a light and the availability of food, like that between the sound of the bell and the satiation of hunger, is artifically contrived. The occurrence of the bell stimulus, however, causes the dog to salivate, whether it wants to or not, whereas the occurrence of a green light does not cause a pigeon to press the bar, whether it wants to or not, but rather establishes a conventional signal for the pigeon that, if it were to perform a bar press now, a pellet would be emitted. It could be argued that the bell stimulus has now become a sufficient condition for the dog to salivate, while the light stimulus has become a sufficient condition for the pigeon not to press the bar. But Skinner's experiments, unlike those of Pavlov, involve reinforcing behavior after it has been performed, because of which Skinner's pigeons, but not Pavlov's dogs, learn means/ends relations over which they have a certain degree of control.

A fascinating example of type/token recognition that displays behavior that appears to be distinctive of semiotic systems of Type I, finally, is described in a newspaper article entitled, "Fake Owls Chase Away Pests" (*St. Petersburg Times*, 27 January 1986), which I was happy to discover:

> Birds may fly over the rainbow, but until 10 days ago, many of them chose to roost on top of a billboard that hangs over Bill Allen's used car lot on Drew Street in Clearwater. Allen said he tried everything he could think of to scare away the birds, but still they came – some times as many as 100 at a time. He said an employee had to wash the used cars at Royal Auto Sales every day to clean off the birds' droppings. About a month ago, Allen said, he called the billboard's owner for help fighting the birds. Shortly afterward, Allen said, two vinyl owl "look alikes" were put on the corners of the billboard. "I haven't had a bird land up there since", he said.

The birds, in other words, took the sizes and shapes of those vinyl owls to be instances of the sizes and shapes of real owls, treating the fake owls as though they were the real thing. Once again, therefore, we can infer that these systems have the capacity to take something to stand for something (else) in some respect or other on the basis of the criterion that they have the capacity to make a mistake, which has been illustrated by Pavlov's dogs, by Skinner's pigeons, and by Allen's birds alike. While there do seem to be criteria distinctive of each of these three types of semiotic systems, in other words, these more specific criteria themselves are consistent with and tend to illuminate that more general criterion, which indicates mentality per se.

These considerations thus afford a foundation for pursuing a comparison of the semiotic approach with the account provided by the symbol system conception. Indeed, the initial conjecture that we are about to explore is that Newell and Simon seem to be preoccupied exclusively with systems of Type III (or special counterparts), which, if true, establishes a sufficient condition for denying that semiotic systems and symbol systems are the same – even while affirming that they are both physical systems (of one or another of the same general kind). The intriguing issues that confront us here, thererfore, concern (a) whether there is any significant difference between semiotic systems of Type III and physical symbol systems, and (b) whether there are any significant reasons for preferring one or another of these accounts with respect to the foundations of artificial intelligence.

Symbol and Causal Systems. The distinction between types and tokens ought to be clear enough by now to consider the difference between Newell and Simon's physical symbol systems and semiotic systems of Type III. The capacity to utilize indices seems to carry with it the capacity to utilize icons, since recognizing instances of causes as events of the same kind with respect to some class of effects entails drawing distinctions on the basis of relations of resemblance. Similarly, the capacity to utilize symbols appears to carry with it the capacity to utilize indices, at least to the extent to which the use of specific symbols on specific occasions can affect the behavior of a semiotic system for which they are significant signs. Insofar as these considerations suggest that a physical symbol system ought to be a powerful kind of semiotic system with the capacity to utilize icons, indices, and symbols as well, it may come as some surprise that I want to deny that Newell and Simon's conception supports such a conclusion at all. For, appearances to the contrary notwithstanding, physical symbol systems in the sense of Newell and Simon do not qualify as systems that use symbols in the sense of semiotic systems.

Because I take it to be obvious that physical symbol systems are causal systems in the appropriate sense, the burden of my position falls upon the distinction between systems for which something functions as a sign for a user of that system and systems for which something functions as a sign for that system itself. Newell and Simon's basic conception is as follows:

> A physical symbol system consists of a set of entities, called symbols, which are physical patterns that can occur as components of another type of entity called an expression (or symbol structure). Thus a symbol structure is composed of a number of instances (or tokens) of symbols related in some physical way (such as one token being next to another). [Newell and Simon (1976), p. 40]

Notice that symbol structures (or "expressions") are composed of sequences of symbols (or "tokens"), where *physical symbol systems*, in this sense, process the expressions Newell and Simon refer to as "symbol structures". The question that needs to be pursued, therefore, is whether or not these "symbol structures" can function as signs in Peirce's sense – and, if so, for whom.

At first glance, this passage may seem to support the conception of physical symbol systems as semiotic systems, since Newell and Simon appeal to tokens, and tokens appear to be instances of different types. Their conceptions of designation and of interpretation, however, are of relevance here:

> Two notions are central to this structure of expressions, symbols, and objects: designation and interpretation.
>
> Designation. An expression designates an object if, given the expression, the system can either affect the object itself or behave in ways depending on the object.
>
> In either case, access to the object via the expression has been obtained, which is the essence of designation.
>
> Interpretation. The system can interpret an expression if the expression designates a process and if, given the expression, the system can carry out the process.
>
> Interpretation implies a special form of dependent action: given an expression, the system can perform the indicated process, which is to say, it can evoke and execute its own processes from expressions that designate them. [Newell and Simon (1976), pp. 40-41]

An appropriate illustration of "interpretation" in this sense would appear to be *computer commands*, where a suitably programmed machine can evoke and execute its own internal processes when given "expressions" that designate them. Notice, on this interpretation, that portable typewriters, pocket calculators, and the like, qualify as physical symbol systems in Newell and Simon's sense. The combinations of letters from their keyboards (of numerals from their interface, and so on) appear to be examples of expressions that designate a process whereby various shapes can be typed upon a page (strings of numerals can be manipulated, and so forth). No doubt, simple systems like these are not the kind that Newell and Simon had in mind as the best examples of symbol systems in their sense. Their focus, after all, is upon general-purpose digital computers. We shall return to this issue.

A consistent interpretation of Newell and Simon's conception depends upon an analysis of "symbols" as members of an alphabet or character set (such as "a", "b", "c", ...), where "expressions" are sequences of members of such a set. The term that they employ which corresponds most closely to that of "symbol" in Peirce's technical sense, therefore, is not "symbol" itself but rather "expression". Indeed, that "symbols" in Newell and Simon's sense cannot be "symbols" in Peirce's technical sense follows from the fact that most of the members of a character set do not stand for (or "mean") anything at all. Their inclusion within such a set simply serves to render them permissible members of the character sequences that constitute the grammatically well-formed expressions of the systems that employ them. A more descriptive name for systems of this kind, therefore, might be that of "expression processing" (or of "string manipulating") systems. So long as the meaning of "symbol" in Newell and Simon's sense is not confused with Peirce's sense, however, it is not necessary to challenge their terminology.

An important consequence of this difference turns out to be that words that occur in ordinary language, like "chair" and "horse", are good examples of symbols in Peirce's sense, yet do not satisfy Newell and Simon's definition of expressions. These signs stand for that for which they stand without in any fashion offering the least hint that the humans (machines or whatever) for which they might possess significance can either affect those objects or behave in ways that depend upon those objects. The capacity to use the word "horse" correctly does not entail the ability to ride or to train them; the capacity to employ the word "chair" properly does not imply the ability to build or to refinish them; and so forth. These symbols can function as significant signs even when Newell and Simon's conditions are not satisfied. A proof of this follows from examples such as "elf" and "werewolf", which function as symbols in Peirce's sense, yet could not possibly fulfill Newell and Simon's conception, because they stand for things that do not exist, which can neither affect nor be affected by causal relations in space/time. Thus, "symbol systems" in Newell and Simon's sense of string manipulating systems do not qualify as systems that utilize "symbols" in Peirce's sense.

Symbol and Semiotic Systems. This result tends to reflect the fact that Newell and Simon's account depends upon at least these two assumptions:

(a) expressions =df sequences of characters (strings of symbols); and,

(b) symbols =df elements of expressions (tokens of character types);

where these "character types" are those specified by some set of characters (ASCII, EBCDIC, . . .). This construction receives further support from other remarks they make during the course of their analysis of completeness and closure as properties of systems of this kind, insofar as they maintain that:

(i) there exist expressions that designate every process of which such a system (machine) is capable; and,
(ii) there exist processes for creating any expression and for modifying any expression in arbitrary ways;

which helps to elucidate the sense in which a system can affect an object itself or behave in ways depending on that object, namely: when that object itself is either a computer command or a string of characters from such a set. [A rather similar conception can be found in Newell (1973), esp. pp. 27-28.]

Conditions (i) and (ii) appear to have the tendency to restrict the class of physical symbol systems to digital computing machines, even though (i) and (ii) can only be satisfied relative to some (presupposed) set of symbols, no matter how arbitrarily selected. Whether or not these conditions actually have their intended effect, however, appears to be subject to debate. While programming languages vary somewhat on this issue, such practices as the overloading of operators, the multiple definition of identifiers, and the like, although overtly encouraged, are covertly ruled out. Specific instructions conveyed to a machine in high-level languages must be unambiguous, in principle, precisely because the machine is engineered to guarantee that "a given instruction elicits a unique operation", as Margaret Boden has observed [Boden (1988), p. 247]. Ambiguous practices are thus important to avoid, which restricts the extent to which the syntax of languages suitable for employment with machines can be arbitrarily composed. Moreover, since typewriters and calculators seem to have unlimited capacities to process any number of expressions that can be formulated within their respective sets of symbols, it seems to follow that they are not excluded by these constraints, which is a striking result, since few of us would be inclined to conjecture that a typewriter, say, has something like "a mind of its own".

No doubt, physical symbol systems in Newell and Simon's sense typically do behave in ways that depend upon certain members of the class of expressions, since they are causal systems that respond to specific computer commands for which they have been programmed. Since computer commands function as input causes in relation to output effects (for suitably programmed machines), it should be obvious that Newell and Simon's conception entails the result that physical symbol systems are causal systems. For reasons indicated above, however, it should be equally apparent that their conception does not entail that these causal systems are semiotic systems of Type III. Indeed, if expressions were symbols in Peirce's technical sense, then they would have to have both intensions and extensions; but it seems clear that strings of symbols from a character set, even when well-formed, need not have these properties. Yet they can still be signs for their users.

Indeed, the most telling considerations of all emerge from inquiring *for whom* Newell and Simon's "symbols" and "expressions" are supposed to be significant (apart from the special class of computer commands, where, in fact, it remains to be ascertained whether or not those commands function as signs for those systems). Consider, for example, the following scenarios:

INPUT	*(FOR SYSTEM)*	*OUTPUT*
finger (pushes)	button (causing)	printout of file
match (lights)	fuse (causing)	explosion of device
child (notices)	cloud (causing)	expectation of rain

Fig. 6. Different Cases.

When a finger pushes a button that activates a process, say, leading to a printout, no doubt an input for a causal system has brought about an output. When a match lights a fuse, say, leading to an explosion, that an input for a causal system has brought about an output is not in doubt. And when a child notices a cloud, say, leading to the expectation of rain, no doubt an input for a causal system has brought about an output. Yet surely only the last of these cases is suggestive of the possibility that something stands for something (else) in some respect or other for that system, where that particular thing is a meaningful token for that system (with an intensional dimension) and where that system might be making or have made a mistake.

If these considerations are correct, then we have discovered, first, that the class of causal systems is not coextensive with the class of semiotic systems. Coffee cups and matches, for example, are particular cases of systems in space/time that stand in causal relations to other things, yet surely do not qualify as sign-using systems. Since two words, phrases, or expressions can mean the same thing only if their extensions are the same, causal systems and semiotic systems are not the same thing. We have also discovered, second, that the meaning of "symbol system" is not the same as the meaning of "semiotic system of Type III". General purpose digital machines are causal systems that process expressions, yet do not therefore need to be systems for which things function as signs. Since two words, phrases, or expressions mean the same thing only if their intensions are the same, symbol systems and semiotic systems of Type III are not the same specific kinds of things.

From the perspective of the semiotic approach, in other words, the conception of physical symbol systems encounters the distinction between sets of symbols that are significant for users of machines – in which case there is a semiotic relationship between those signs, what they stand for and those sign users, where the users are *not* identical with the machines themselves – and sets of symbols that are significant for use by machines – in which case there is a semiotic relationship between those signs, what they stand for, and those machines, where these users *are* identical with the machines themselves. The position I am drawing here can be readily diagrammed as follows:

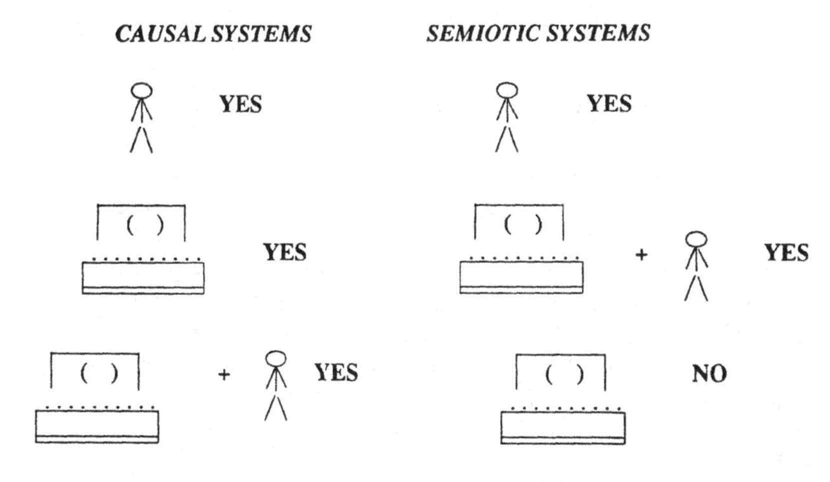

Fig. 7. The Big Picture.

Without any doubt, the symbols and expressions with which programmers program machines are significant signs for those programmers (although the same program may not always make sense to some other programmer). Without any doubt, the capacity to execute those commands qualifies those commands as causal inputs with respect to causal outputs. But that is not enough for any digital machines to qualify as semiotic systems of Type III.

If these reflections are well-founded, there is a fundamental difference between causal systems and semiotic systems, on the one hand, and between symbol systems and semiotic systems of Type III, on the other. Of course, important questions remain, including ascertaining whether or not there are good reasons for preferring one or another conception as the foundation for AI. Moreover, there appear to be several unexamined alternatives with respect to Newell and Simon's conception, since other arguments might be advanced to establish that symbol systems properly qualify either as semiotic systems of Type I or of Type II or else that special kinds of symbol systems properly qualify as semiotic systems of Type III, which would seem to be an important possibility that has yet to be considered. For the discovery that some symbol systems are not semiotic systems of Type III no more proves that special kinds of symbol systems cannot be semiotic systems of Type III than the discovery that some causal systems are not semiotic systems proves that special kinds of causal systems cannot be semiotic systems.

THE SYMBOL AND SEMIOTIC-SYSTEM HYPOTHESES

The conception of semiotic systems, no less than the conception of symbol systems, can be evaluated (at least in part) by the contributions they make toward illuminating the relationship between the use of signs by semiotic systems or the manipulation of strings by symbol systems and the production of purposive behavior. These accounts, in other words, can be viewed as offering characterizations of "mental activity" as alternative conjectures, where both of their conceptions are intended to afford a basis for comprehending "intelligent" (or "deliberate") behavior. The respective theoretical hypotheses that they represent, moreover, may be formulated as follows:

(*h*1) *The Symbol-System Hypothesis*: a symbol system has the necessary and sufficient means (or capacity) for general intelligent action; and,
(*h*2) *The Semiotic-System Hypothesis*: a semiotic system has the necessary and sufficient means (or capacity) for general intelligent action.

Both of these hypotheses are to be entertained as empirical generalizations (or as lawlike claims) whose truth and falsity cannot be ascertained merely by reflection upon their meaning within a certain language framework alone.

Since both hypotheses propose necessary and sufficient conditions, they could be shown to be false if either (a) systems that display "intelligent" (or "deliberate") behavior are not symbol (or semiotic) systems, or (b) systems that are symbol (or semiotic) systems do not display "intelligent" (or "deliberate") behavior. Moreover, since they are intended to be empirical conjectures [cf. Newell and Simon (1976), esp. p. 42 and p. 46], these formulations ought to be understood as satisfied by systems that display appropriate behavior without assuming (i) that behavior that involves the processing or the manipulation of strings of tokens from a character set is therefore either "intelligent" or "deliberate" (since otherwise the symbol-system hypothesis must be true as a function of its meaning), and without assuming (ii) that behavior that is "intelligent" or "deliberate" must therefore be successful in attaining its aims, objectives, or goals, where a system of this kind cannot make a mistake (since otherwise the semiotic-system hypothesis must be false as a function of its meaning). With respect to the hypotheses (*h*1) and (*h*2) before us, therefore, "intelligent action" and "deliberate behavior" are to be treated as though they were synonymous expressions.

A certain amount of vagueness inevitably attends an investigation of this kind to the extent to which the notions upon which it depends, such as "deliberate behavior" and "intelligent action", are not fully defined. Nevertheless, an evaluation of the relative strengths and weaknesses of these hypotheses can result from considering classes of cases that fall within the extensions of "symbol system" and of "semiotic system" when properly understood. In particular, it seems obvious that the examples of type/token recognition, of classical conditioning, and of instrumental conditioning that we have considered above are instances of semiotic systems of Type I, II, and III that do not qualify as symbol systems in Newell and Simon's sense. That this is the case should come as no surprise, since (almost certainly) Newell and Simon did not intend that their conception should apply with such broad scope; but it evidently entails the result that hypothesis (*h*1) must be *empirically false*.

Indeed, while this evidence amply supports the conclusion that the semiotic-system approach has applicability to dogs, to pigeons, and to other birds that the symbol-system approach lacks, the importance that ought to attend this realization might or might not be immediately apparent. Consider, after all, that a similar argument could be made on behalf of the alternative account, namely: that the symbol-system definition has applicability to typewriters, to calculators, and to other machines that the semiotic-system conception lacks, which may be of even greater importance if the objects that are of primary interest happen to be inanimate machines. For if digital computers, for example, have to be symbol systems but do not have to be semiotic systems, then that they do not qualify as semiotic systems is not necessarily a matter of immense theoretical significance. The importance of the answer depends upon the importance of different questions one might ask.

Nevertheless, to the extent to which these respective conceptions are intended to have the capacity to shed light on the general character of the causal connections that obtain between mental activity and behavioral tendencies – that is, to the extent to which frameworks like these ought to be evaluated in relation to hypotheses such as (*h*1) and (*h*2) – the evidence that has been assembled here would appear to support the conclusion that the semiotic-system approach clarifies connections between mental activity as semiotic activity and behavioral tendencies as deliberate behavior – connections which, by virtue of its restricted range of applicability, the symbol-system approach cannot accommodate. By combining distinctions between different kinds (or types) of mental activity together with psychological criteria concerning the sorts of capacities distinctive of systems of these different kinds (or types), the semiotic approach provides a powerful combination of (explanatory and predictive) principles, an account that, at least in relation to human and non-human animals, the symbol-system hypothesis cannot begin to rival. From this point of view, the semiotic-system conception, but not the symbol-system conception, appears to qualify as a theory of mind.

What about Humans and Machines? Perhaps the most defensible response that Newell and Simon might embrace in the face of these results would be to abandon their commitment to the symbol-system hypothesis and restrict the scope of their analysis to the thesis that general-purpose digital machines are systems of the kind they have defined. Then it really doesn't matter whether or not they have successfully captured the nature of mental activity in humans or in other animals. In this case, it would be sensible to acknowledge the real possibility that their analysis may have captured no sense of mentality at all. But that would not mean that they have failed to diagnose the nature of at least some "intelligent machines", especially when these are understood to be artificial contrivances which have the ability to perform certain types of tasks with enormous success.

The only alternative would be to contend that the processes that characterize symbol systems *are* the same processes that characterize other animals and human beings, which is not consistent with the analysis we have pursued. Notice, in particular, that the following theses concerning the relationship of systems of these two kinds may both be true, namely:

(*t*1) general purpose computing machines are symbol systems; and,

(*t*2) animals – human and non-human alike – are semiotic systems.

In fact, even if no digital computer heretofore constructed qualifies as a semiotic system, that some computing machine yet to be built might later qualify as a semiotic system remains an open question. The question of whether knowledge, information, and data processing in humans and in other animals is more like that of symbol systems or more like that of semiotic systems can thus be viewed as the fundamental question of AI.

Artificial intelligence is frequently taken to be an attempt to develop causal processses that perform mental operations. Sometimes this view has been advanced in terms of *formal systems*, where "intelligent beings are . . . automatic formal systems with interpretations under which they consistently make sense" [Haugeland (1981), p. 31]. This conception exerts considerable appeal, because it offers the prospect of reconciling a domain about which a great deal is known – formal systems – with one about which a great deal is not known – intelligent beings. This theory suffers from profound ambiguity, however, since it fails to distinguish between systems that make sense to themselves and those that make sense for others. Causal models of mental processes, after all, might either affect connections between inputs and outputs so that, for a system of a certain specific type, those models yield outputs for certain classes of inputs that correspond to those exemplified by the systems that they model, or else affect those connections between inputs and outputs and, in addition, process these connections by means of processes that correspond to those of the systems that they model.

This difference, which is already familiar, can be specified by distinguishing between "simulation" and "replication" as previously proposed:

(a) Causal models that *simulate* mental processes capture connections between inputs and outputs that correspond to those of the systems that they represent.

(b) Causal models that *replicate* mental processes not only capture the connections between inputs and outputs but do so by means of processes that correspond to those of the systems that they represent.

Thus, if theses (*t*1) and (*t*2) *are* true, then it may be said that symbol systems simulate mental processes that semiotic systems replicate – precisely because semiotic systems have minds that symbol systems lack.

There are those, such as Haugeland (1985), of course, who are inclined to believe that symbol systems replicate mental activity in humans as well because human mental activity, properly understood, has the properties of symbol systems, too. But this claim appears to be plausible only if there is no real difference between systems for which signs function as signs for those systems themselves and systems for which signs function as signs for the users of those systems, which is the issue in dispute. My analysis implies that he is wrong.

Another perspective on this matter can be secured by considering the conception of systems that possess the capacity to represent and to utilize information or knowledge. The instances of semiotic systems of Types I, II, and III that we have explored seem to fulfill this desideratum, in the sense that, for Pavlov's dogs, for Skinner's pigeons, and for Allen's birds, there are clear senses in which these causal systems are behaving in accordance with their beliefs, that is, with something that might be properly characterized as "information" or as "knowledge". Indeed, the approach presented here affords a welcome opportunity to relate genes to bodies to minds to behavior, since phenotypes develop from genotypes under the influence of environmental factors, where phenotypes of different kinds may be described as predisposed toward the utilization of different kinds of signs. The various species are inclined to the acquisition and utilization of distinct ranges of behavioral tendencies, which have their own distinctive strengths [Fetzer (1985b), (1988a)]. This in itself offers significant incentives for adopting the semiotic approach toward understanding mind.

Yet it could still be the case that digital computers (pocket calculators, and the like) cannot be subsumed under the semiotic framework, precisely because theses (*t*1) and (*t*2) are *both* true. After all, nothing that has gone before alters obvious differences between systems of these various kinds, which are created or produced by distinctive kinds of causal processes. The behavior of machine systems is highly artificially determined or engineered, while that of human systems is highly culturally determined or engineered, and that of other animal systems is highly genetically determined or engineered. Systems of all three kinds exhibit different kinds of causal capabilities: they differ with respect to their ranges of inputs/stimuli/trials, with respect to their ranges of outputs/responses/outcomes, and with respect to their higher-order causal capabilities, where humans (among animals) appear superior. So even if theses (*t*1) and (*t*2) are both true, what difference does it make?

What Difference Does It Make? From the point of view of the discipline of artificial intelligence, whether computing machines do what they do the same way that humans and other animals do what they do only matters in relation to whether the enterprise is supposed to be that of simulating or of replicating the mental processes of semiotic systems. If the objective is simulation, it is surely unnecessary to develop the capacity to manufacture semiotic systems; but if the objective is replication, there is no other way, since this aim cannot otherwise be attained. Yet it seems to be worth asking whether the replication of the mental processes of human beings should be worth the time, expense, and effort that would be involved in building them. After all, we already know how to reproduce causal systems that possess the mental processes of human beings in ways that are cheaper, faster, and lots more fun. Indeed, when consideration is given to the limited and fallible memories, the emotional and distorted reasoning, and the inconsistent attitudes and beliefs that tend to distinguish systems of this kind, it is hard to imagine why anyone would want to build them. There are *no* interpretations "under which they consistently make sense"!

A completely different line could be advanced by defenders of the faith, however, who might insist that the distinction I have drawn between symbol systems and semiotic systems cannot be sustained, because the conception of semiotic systems itself is circular and therefore unacceptable. If this contention were correct, the replication approach might be said to have been vindicated by default in the absence of any serious alternatives. The basis for this objection could be rooted in a careful reading of the account that I have given for semiotic systems of Type I, since, within the domain of semantics, icons are supposed to stand for that for which they stand when they create in the mind of a sign user another – equivalent or more developed – sign that stands in the same relation to that for which they stand as do the original signs creating them. This Peircean point, after all, employs the notion of mind in the definition of one of the kinds of signs – the most basic kind, if the use of indices involves the use of icons and the use of symbols involves the use of indices, but not conversely – which might be thought to undermine any theory of the nature of mind based upon Peirce's theory of signs.

This complaint, I am afraid, is founded upon an illusion. For those signs in terms of which other signs are ultimately to be understood are unpacked by Peirce in terms of the habits, dispositions, or tendencies by means of which all signs are best understood: "the most perfect account of a concept that words can convey", he wrote, "will consist of a description of the habit which that concept is calculated to produce" [Hartshorne and Weiss (1960), 5.491]. Yet this result in turn could provide another avenue of defense by contending that systems of dispositions cannot be causal systems, so that, *a fortiori*, semiotic systems cannot be causal systems as part of a dispositional account. Without suggesting that the last word has been said with reference to this question – which is explored in Chapter 3 – there seems to be no evidence in its support; but it would defeat the analysis I have presented if such an argument were correct.

The basic distinction between symbol systems and semiotic systems, of course, is that symbol systems do not seem to be systems for which signs can stand for something (else) in some respect, while semiotic systems are. I have employed the general criterion that semiotic systems are capable of making mistakes. A severe test of this conception, therefore, is raised by the problem of whether or not digital computers, especially, are capable of making a mistake. If the allegations that the supercomputers of the North American Defense Command which is located at Colorado Springs have reported the U.S. to be under ballistic missile attack from the Soviet Union no less than 187 times are (even remotely) accurate, this dividing line may already have been crossed, since it appears as though all such reports thus far have been false. The problem, however, is deciding whether these sophisticated computing systems themselves infer that such attacks are taking place or merely report data that is subject to misinterpretation by their human users. The systems most promising for making their own mistakes, if any were to turn out to possess this capability, moreover, are those for which a faulty inference can occur, the false can be mistaken for the true, or things can be misclassified, which might not require systems more complex than those capable of playing chess [Haugeland (1981), p. 18]. However, this question, we have discovered, is heavily loaded theoretically.

The evidence that has been offered in support of the semiotic conception, after all, appears to acquire its persuasive appeal by virtue of its capacity to clarify and illuminate a broad range of issues that fall within the domain of understanding the mind. Individually, they may seem vulnerable to rebuttal, but collectively, they appear to be mutually reinforcing. In the case of Skinner's pigeons, for example, it might be alleged that, when the apparatus has been altered, the pigeons do not err but rather the conditions have simply changed. By the same token, of course, the target of assassination who turned his ignition key and blew himself to pieces would not have made a mistake either. Our reasoning may be inconclusive in matters of this kind, but that should not render us incapable of appreciating an elegant theory.

As semiotic systems, human beings display certain higher-order causal capabilities that deserve to be acknowledged, since we appear to have a remarkable capacity for inferential reasoning that may or may not differ from that of other animals in kind but undoubtedly exceeds them in degree. In this respect, especially, human mental abilities are themselves surpassed by the operational performance of "reasoning machines", which are, in general, more precise, less emotional, and far faster in arriving at conclusions by means of deductive inference. The position defended here, of course, itself entails that computers cannot add, subtract, etc., when that is taken to imply *understanding* the nature of addition, of subtraction, etc. [cf. Dretske (1985)]. Nevertheless, even if symbol systems are incapable of mentality, the evolution and development of newer systems with inductive and perceptual capabilities promises to be the most likely source of devices that might have the capacity to make mistakes. By this criterion, after all, systems that have the capacity to make mistakes qualify as semiotic systems, even when they do not otherwise replicate the processes of human beings.

A form of mentality that exceeds the use of symbols alone, moreover, appears to be the capacity to make assertions, to issue directives, to ask questions, and to utter exclamations. At this juncture, I think, the theory of minds as semiotic systems intersects with the theory of languages as transformational grammars, especially as it is found in the work of Noam Chomsky [Chomsky (1965), (1966), for example; but see Chomsky (1986) for his more recent views]. Not much should be made of this coincidence, however, since I embrace a dispositional framework that Chomsky would no doubt reject. This connection, however, suggests the possibility that it might be desirable to identify a fourth grade of mentality, where semiotic systems of Type IV can utilize signs that are transformations of other signs.

An evidential indicator of mentality of this type would appear to be the ability to ask questions, make assertions, and the like. The capacity for logical reasoning, moreover, is an extremely important instance of this general ability. Logical reasoning involves argumentation in the sense of providing reasons for action or for belief, especially with respect to explanations and predictions. While all kinds of thinking appear to involve the use of signs, therefore, not all kinds of thinking involve reasoning. Contrary to the computational conception, which we shall pursue in Chapter

3, it appears to be the theory of signs that defines the boundaries of thought. This suggests that the ability to reason logically can serve as a criterion of mentality of Type IV, provided that it is understood as a special ability to offer reasons.

Humans, other animals, and machines, of course, also seem to differ with respect to other higher-order mental capabilities, such as in their attitudes toward and beliefs about the world, themselves, and their methods. Indeed, I am inclined to believe that those features of mental activity that separate humans from other animals occur at just this juncture. For humans have a capacity to examine and to criticize their attitudes, their beliefs, and their methods that other animals do not appear to enjoy. From this perspective, however, the semiotic approach seemingly should classify symbol systems as engaged in a species of activity – involving the manipulation of strings – that, were it pursued by human beings, would occur at this level. For the activities of linguists, of logicians, and of critics in creating and processing expressions and symbols certainly appear to be higher-order activities.

A fifth grade of mentality thus deserves to be acknowledged as a mode of meta-mentality that distinguishes itself by the use of signs to stand for other signs. While semiotic systems Type I can utilize icons, of Type II indices, of Type III symbols, and of Type IV transforms, semiotic systems of Type V can use meta-signs as signs that stand for other signs (one

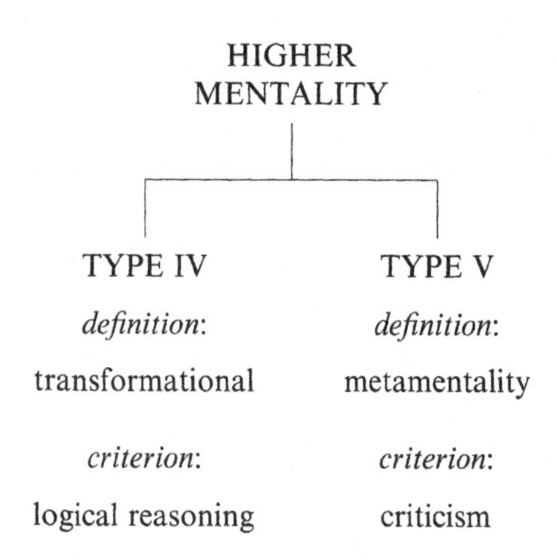

Fig. 8. Higher Types of Mentality.

especially important variety of meta-signs, of course, being meta-languages, i.e., languages that can be used to talk about other languages). Thus, perhaps the crucial criterion of mentality of this degree is the capacity for criticism, of ourselves, our theories, and our methods. While the conception of minds as semiotic systems has a deflationary effect in rendering the existence of mind at once more ubiquitous and less important than we have heretofore supposed, it does not therefore diminish the place of human minds as semiotic systems of a distinctive kind, nevertheless.

The introduction of semiotic systems of Type IV and Type V, however, should not be allowed to obscure the three most fundamental species of mentality. Both logical reasoning and critcial capacities appear to be varieties of semiotic capability that fall within the scope of symbolic mentality. Indeed, as a conjecture, it appears to be plausible to suppose that each of these successively higher and higher types of mentality presupposes the capacity for each of those below, where evolutionary considerations might be brought to bear upon the assessment of this hypothesis by attempting to evaluate the potential benefits for survival and reproduction to species and societies – that is, for social groups as well as for single individuals – that accompany this conception [Fetzer (1990)].

If the arguments developed in this chapter are well-founded, then the presumption that physical symbol systems provide an appropriate foundation for understanding human mental processes is an unjustifiable inference. Indeed, while human beings and digital machines may qualify as causal systems, they appear to function in significantly different ways. Digital machines as symbol systems have the ability to manipulate strings of characters even though those strings do not stand for anything at all for systems of that kind. Human beings as semiotic systems, by comparison, have the ability to manipulate strings of characters which often stand for other things for systems of that kind. The belief that the symbol system conception supplies the elements required to replicate human mentality thus depends on a faulty analogy between systems of two different kinds.

There remains the further possibility that the distinction between symbol systems and semiotic systems marks the dividing line between computer science (narrowly defined) and artificial intelligence, which is not to deny that artificial intelligence falls within computer science (broadly defined). On this view, what is most important about artificial intelligence as an area of specialization within this field itself would be its ultimate objective of replicating semiotic systems. Indeed, while artificial intelligence can achieve at least some of its goals by building systems that simulate – and improve upon – the mental abilities that are displayed by human beings, it cannot secure its most treasured goals short of replication, if such a conception is correct. It therefore appears to be an ultimate irony that the ideal limit and final aim of artificial intelligence could turn out to be the development of systems capable of making mistakes.

3. THEORIES OF LANGUAGE AND MENTALITY

There are several reasons why the nature of language is fundamental to research in artificial intelligence in particular and to cognitive inquiry in general. One tends to be the assumption – considered in part in Chapter 2 – that thinking takes place in language, which makes the nature of language fundamental to the nature of mental processes, if not to the nature of mind itself. Another is that computers operate by means of software composed by means of a language – not a natural language, to be sure, but a computer language, which is a special kind of artificial language suitable for conveying instructions to machines. And another is that debate continues to rage over whether or not machines can have minds, a question whose answer directly depends upon the nature of mentality itself and indirectly upon the nature of language, especially the nature of languages suitable for use by machines.

Below the surface of these difficulties, however, lies another problematic question, namely: is artificial intelligence *descriptive* or *normative*? For if artificial intelligence is supposed to utilize the methods that human beings themselves – descriptively – actually employ in problem solving, then there would appear to be a powerful motive for insuring that the languages used by machines are similar (in all relevant respects) to those used by humans. If artificial intelligence is not restricted to the methods that human beings actually use but implements those that humans ought to use – prescriptively – whether or not they actually do, then whether computer languages are like or unlike natural languages at once appears to be a less pressing issue.

Most students of artificial intelligence tend to fall into two broad (but heterogeneous) camps. One camp recommends the "strong" thesis that AI concerns how we do think. The other embraces the "weak" thesis that AI concerns how we ought to think. And there are grounds to believe that the predominant view among research workers today is that the strong thesis is correct. Since Charniak and McDermott want to create artificial thinking things whose processes replicate those of human beings, it is evident that they are supporters of strong AI. An assumption underlying this point of view is that, at some level, the way in which the mind functions is the same as the way in which some kinds of computational systems (digital machines, especially) also function. This assumption, however, is one that adherents of both camps might endorse, insofar as even normative approaches to AI presumably would need to satisfy this condition "at *some* (suitable) level".

An alternative conception – yet a third approach – would be to envision AI as concerning how we could think, whether that is how we do think or how we should think. The distinction between "do", "should", and "could" modes of operation may be less great than it intially appears. Modes that we do use are obviously modes we could use, since otherwise it would not be possible for us to use them. Modes that we should use, moreover, are also modes we could use, at least in relation to the normal assumption of ethical theory that "ought" implies "can". Indeed, even this perspective similarly appears to presuppose that "at *some* (suitable) level" the modes of operation of human minds and of computational systems are the same.

Nevertheless, the greatest benefit of this further point of view may be its heuristic power in strongly suggesting that any modes of mental operation compatible with the underlying architecture deserve to be considered. In fact, since the relationship of compatibility between "architecture" and "mode of operation" must be *symmetrical*, it tacitly invites exploration of different accounts of the underlying architecture itself. Various models of of the brain (including, of course, connectionist conceptions) might afford a better framework for understanding the mind – provided, of course, that they are compatible with possible modes of operation of human minds. It indicates that models of the mind might be adapted to models of the brain or that models of the brain might be adapted to models of the mind instead.

One of the ways in which the "strong" position can be infused with content arises from a more specific characterization of the "internal representations" that may attend the conception previously elaborated in Chapter 1. Charniak and McDermott, in particular, offer a depiction of a program that receives information as input and answers questions as output, a process that proceeds through four distinct steps or stages more or less as follows:

Step 1: when the program gets a statement, it translates it into an internal representation and stores it away;

Step 2: when it gets a question, it translates it into an internal representation as well;

Step 3: it uses the internal representation of the question to fetch information from its memory;

Step 4: it translates the answer back into English (or some other interactive language). [Charniak and McDermott (1985), pp. 11-12]

Notice, however, that this description might be intended *literally* ("This is how the machine does it") or *figuratively* ("The machine behaves as if it were doing it this way"), which corresponds to the distinction between simulation and replication explained above. Simulations might therefore be said to involve figurative representations (from an "external" point of view), whereas replications involve literal representations (from an "internal" point of view) – assuming that this is something that humans can do.

This general approach, moreover, has been reinforced by the proposition that, when mental processes are viewed as computational, minds themselves can be viewed as special kinds of formal systems. Haugeland (1981, 1985), as we have observed, has gone so far as to suggest that mental activity can be adequately portrayed as the behavior of an *automated formal system*, a position that leads him to the conjecture, "Why not suppose that people just *are* computers (and send philosophy packing)?" [Haugeland (1981), p. 5]. Indeed, the prospect of reducing the philosophy of mind to problems of design in the construction and development of digital machines has excited a host of adherents across many fields and disciplines, including those of linguistics, psychology, and philosophy, as well as those of computer science and AI.

If the internal representations of a machine can be utilized in storing and fetching information from its memory, then the internal representations of human beings might possibly function in similar ways – assuming, once again, that some version of the computational requirement could be satisfied. The question that arises at this juncture thus becomes: exactly what characteristics, properties, or features of *human beings* do "internal representations" themselves represent? The answer would appear to be the information, knowledge, or beliefs that are the "internal stuff" that is acquired and modified by the cognitive processes of human beings. Auditory and visual sensations, for example, might be processed into information, knowledge, or beliefs that, in turn, are subject to additional processes of deduction, search, planning, and so forth, yielding tendencies to display physical and verbal behavior under appropriate conditions, where these conditions are similarly processed as further internal representations, etc.

The tenability of this computational conception, however, has not gone unchallenged. Indeed, from a perfectly general perspective rooted in what is known as *semiotic* (explored in Chapter 2), there are three fundamental aspects to systems of signs generally and to languages specifically. Notice, after all, that if there are three dimensions of signs, how could a theory of language that focuses on only one be expected to provide an adequate account of language and mentality? The analysis of Chapter 2, moreover, is focused upon the comparative merits of two specific conceptions of the nature of mental activity, where the symbol-system proposal reflects merely one within a spectrum of possible computational accounts. It has not been shown that no other computational approach could successfully displace it.

My purpose in this chapter, pursuing this lead, is to explore three alternative frameworks for understanding the nature of language and mentality, which accent syntactical, semantical, and pragmatical dimensions of the phenomena with which they are concerned, respectively. Although the computational conception currently exercises considerable appeal, its defensibility seems to hinge upon an extremely implausible theory of the relation of form to content. Similarly, while the representational approach has much to recommend it, its range is essentially restricted to those units of language that can be understood in terms of undefined units. Hence, the only alternative among these three that seems capable of accounting for the meaning of the primitive units of language emphasizes the role of skills, habits, and tendencies relating signs and dispositions. And the same considerations provide a foundation for assessing conceptions of these kinds as "models of the mind".

Perhaps one cautionary note is in order before pursing this objective. The methodology being employed throughout is analytical rather than historical, in the sense that the subject of interest is *the problem space* – or, even better, *the solution space* – appropriate to these problems. I am less concerned with the detailed positions that have been held by the specific individuals – such as Haugeland, Fodor, Stich, and others – whose works are mentioned in passing than I am with the general features of the problems and solutions toward which they are directed. By emphasizing the predominantly syntactical, semantical, and pragmatical aspects of the alternatives considered, their essential dimensions may be perceived more clearly and their relative plausiblity may be assessed more accurately, a procedure not unlike relating surface phenomena to their deep structure.

THE COMPUTATIONAL CONCEPTION

Computational conceptions of language and of mind depend upon the assumption that languages and mental processes can be completely characterized by means of purely formal distinctions. In discussing the computational conception (or "model") of language and of mind, for example, Fodor has observed that such a construction entails the thesis that "mental processes have access only to formal (non-semantic) properties of the mental representations over which they are defined" [Fodor (1980), p. 307]. Thus,

> the computational theory of the mind requires that two thoughts can be distinct in content only if they can be identified with relations to formally distinct representations. More generally: fix the subject and the relation, and then mental states can be (type) distinct only if the representations which constitute their objects are formally distinct. [Fodor (1980), p. 310]

Notice that at least two issues are intimately intertwined in this passage, for Fodor is maintaining (a) that thoughts are distinct *only if* they can be "identified" with distinct representations, without explaining how it is (b) that specific thoughts can be identified with specific representations. As a necessary condition for thought identity, in other words, condition (b) must be capable of satisfaction as well as condition (a); otherwise, his account will be purely syntactical as an analysis of the relations between signs lacking significance.

Fodor, Stich, and others, too, have explored various ways in which specific "representations" might be identified with specific "thoughts" (in other words, how forms could be infused with content). The strongest versions of the computational conception, however, tend to eschew concern for matters of semantics, as Stephen Stich, for example, has quite lucidly emphasized:

> On the matter of content or semantic properties, the STM [the Syntactic Theory of the Mind] is officially agnostic. It does not insist that syntactic state types have no content, nor does it insist that tokens of syntactic state types have no content. It is simply silent on the whole matter.... the STM is in effect claiming that psychological theories have no need to postulate content or other semantic properties, like truth conditions. [Stich (1983), p. 186]

In order to preserve the differences between the syntactical character of the computational conception and the semantical character of its representational counterpart, we shall defer our consideration of thesis (b) until later on.

Fodor is suggesting an exceptionally strong connection between the *form* of a thought and its *content*. The strongest version of such a position would appear to be that mental tokens (which might be sentences in a natural language, inscriptions in a mental language, or some other variety of types and tokens capable of formal discrimination) have the same content (express the same thought, convey the same idea, or otherwise impart the same information) *if and only if* they have the same form (where "mental tokens" occur as instances of mental types that might have specific content). Differences and similarities, formal or not, of course, presuppose a point of view, which establishes a standard of "sameness" for tokens with respect to whether or not they qualify as "tokens of the same type". Assuming that this condition can be satisfied by systems for which formal distinctions are fundamental, I shall later use a few examples from ordinary English as helpful illustrations.

Any biconditional, of course, is logically equivalent to the conjunction of two conditionals. In this case, those two conditionals are (i) if mental tokens have the same form, then they have the same content, and (ii) if mental tokens have the same content, then they have the same form. Since Fodor has stipulated that contents differ only if they can be identified with different forms, he appears to endorse thesis (i), whose contrapositive maintains that if mental tokens do not have the same content, then they do not have the same form. Although we have observed that Fodor does not elaborate how distinct thoughts are connected to distinct tokens, we shall assume that he intends that, say, similar surface grammars sometimes conceal differences in meaning that are disclosed by an investigation of their deep structures, which means that their parse trees, if they were parsed, would differ, etc.

There would seem to be at least three different ways in which tokens having the same form might be exhibited as possessing different content, each of which appears to be successively less and less syntactical in nature:

(1) *the parsing criterion*, according to which tokens have the same content only if they have the same parse trees;
(2) *the substitutional criterion*, according to which tokens have the same content only if they are mutually derivable by the substitution of definiens for definiendum; and,

(3) *the functional role criterion*, according to which tokens have the same content only if they fulfill the same causal role in their effects upon behavior relative to all possible situations.

Each of these, no doubt, requires unpacking to be clearly understood. Since they accent syntactical, semantical, and behavioral dimensions of language and mentality, in turn, we shall consider them separately in the following, beginning with the parsing criterion in relation to the syntactical approach.

Distinctive versions of the computational conception might defend one of these conditionals but abandon the other, while retaining their computational character; indeed, that appears to be Fodor's own position here. But it is difficult to imagine how a position that abandoned *both* of these theses could still qualify as a "computational" conception. Thus, in order to display the poverty of the computational approach, arguments will be advanced to show that both of these conditionals are *false*. Moreover, the reasons that they are false are surprisingly obvious, but none the less telling. The first of these conditionals falls prey to the problem of ambiguity, while the second succumbs to the problem of synonymy, as numerous examples display.

The Parsing Criterion. The first conditional claims that mental tokens with the same form must have the same content. This thesis would not be true if mental tokens can have the same form yet differ in their content. If ordinary sentences in the English language qualify as "mental tokens", therefore, then examples involving ambiguous words – such as "hot", "fast", etc. – generate counterexamples whose status as ambiguous sentences is not likely to be challenged. Consider the following specific examples as illustrations:

Example (A): Imagine a very warm summer afternoon, as a group of guys are conversing about a shiny red convertible. One casually remarks,

(a) "That car is hot!"

Clearly, at least two meanings (contents) might be intended by remark (a), since it might be interpreted as a comment on the temperature of the car,

(b) "That car has heated to over 100 degrees Farenheit!"

but might be meant to convey its status as a recent and illegal acquisition:

(c) "The cops are out looking for that car everywhere!"

In this case, tokens of the same form, (a), could thus have different content.

Example (B): Sitting on the steps of the stadium, two girls are engaged in animated conversation. As a young guy passes by, one says to the other,

(d) "John is fast!"

Once again, at least two meanings (contents) might be intended by remark (d), since the speaker might be commenting on John's prowess as a sprinter,

(e) "John can beat almost everyone else at the 100 yard dash!"

but it might also be intended as a characterization of his dating behavior,

(f) "John wants to go farther faster than anyone I have dated!"

Once again, therefore, tokens of the same form (d) could differ in meaning.

Initially, ambiguous tokens, such as these, might seem to pose no problem for the computational conception, since the parsing criterion could be employed to discriminate between them. Since the construction of a parse tree presumes the availability of a suitable grammar for this purpose [cf. Winograd (1983)], let us adopt the following very elementary grammar, *G*:

Grammar G:
⟨sentence⟩ → ⟨noun phrase⟩ ⟨verb phrase⟩
⟨noun phrase⟩ → ⟨proper noun⟩
⟨noun phrase⟩ → ⟨determiner⟩ ⟨proper noun⟩
⟨verb phrase⟩ → ⟨copulative verb⟩ ⟨adjective⟩

which, for this exercise, is all of the grammar that happens to be required.

It is important to observe, therefore, that the parse tree for example (a) turns out to have the following structure,

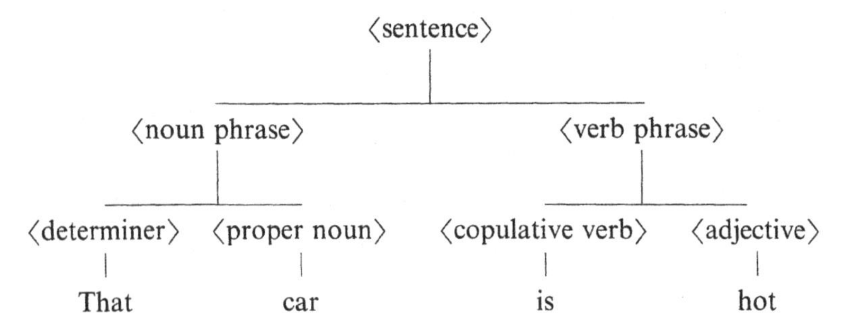

while the parse tree for example (d) has the following different structure,

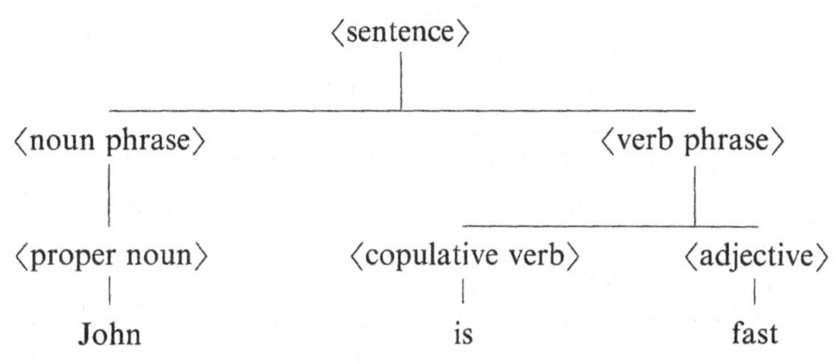

Thus, in terms of the parsing criterion, example (a) and example (d) clearly differ in their parse trees and thus differ in their form, which presumably provides *prima facie* evidence in support of the computational conception. But this is the case only if they also differ in their content. We only know that they differ in their content, because we have assumed they are in English.

The point is that, while it is indeed true that, relative to grammar *G*, the parse trees for sentences (a) and (d) are indeed different, that supports thesis (i), which asserts that tokens with different content have different forms, *only if* sentences (a) and (d) do have different content. We have taken for granted that the sentences, "That car is hot!" and "John is fast!", have different meanings and therefore ought to have different forms, provided thesis (i) is true. If, however, the object referred to by the noun phrases, "that car" and "John", were the same and the properties ascribed by the verb phrases, "is hot" and "is fast", were the same, then the results attained above would count as a counterexample *against* thesis (i) rather than as evidence *for* it. If we did not already understand their content, we could not tell the score.

Ambiguity and Synonymy. Assuming these examples are well-founded, there is no basis for denying that in ordinary English sentence (a) is used to mean what is expressed by sentence (b) but also to mean what is expressed by sentence (c) – and, analogously, for sentence (d) in relation to sentences (e) and (f). Thus, these examples clearly establish that thesis (ii) is false, insofar as there are at least some sentences in English that have the same content but different forms, which would be impossible if mental tokens having the same content have to have the same form. It is evidently not the case that, if mental tokens differ in their form, then they differ in their content. This thesis of computationalism cannot cope with the problem of synonymy.

Now imagine an impoverished version of English, sub-English-4, say, in which the *only* words in the vocabulary happen to be those that are found in sentence (a) – or, alternatively, sub-English-3, those in sentence (d). Naturally, these are fanciful scenarios, since, for a variety of reasons, these are not likely to really be the *only* words in anyone's vocabulary. Nevertheless, they represent a class of cases of the relevant kind. Under these circumstances, would it not be possible that (a) could sometimes be used to mean (b) and sometimes to mean (c) or that (d) could sometimes be used to mean (e) and sometimes to mean (f)? Of course, it would not be possible to *articulate* the contemplated difference. Would that mean *no* such difference could exist?

There are strong reasons for thinking it would not. Consider, after all, that some of the most frequently used exclamations in the English language vary in their meaning with the context of their use: "Damn!", for example, is sometimes used as an expression of distress (discomfort or pain), but it also occurs as an expression of relief (happiness or joy). Here the same word is being used with very different content. Similarly, other criteria, such as the functional role criterion, could differentiate between different meanings for sentence (a): if the gang were to cease talking and scatter when a police car approached, that would tend to indicate that (a) was being used in sense (c); if they cracked an egg on the hood of the car when they wanted to eat it for a snack, that would tend to indicate that (a) was being used in sense (b); etc.

In cases of this kind, these differences in meaning (or content) would indeed exist, even though they could not be articulated *within those languages*. If these examples are acceptable, therefore, then it seems clear that thesis (i) is also false, insofar as at least some sentences in English (sub-English-4 or sub-English-3) have the same form but different contents, which would be impossible if mental tokens having the same form have to have the same content. This thesis of computationalism cannot cope with the problem of ambiguity. Appeals to dependence of meaning upon context, moreover, can be extended to larger classes of sentences through considering successively larger and larger units of discourse. But it should not be overlooked that successively larger and larger units of discourse remain vulerable to problems of ambiguity, to problems of synonymy, and to problems of context.

There are no grounds at all to believe that longer and longer sentences (sequences of symbols, strings of marks) are necessarily or inevitably less and less ambiguous, as the computational conception would suggest. Sets of sentences constituting paragraphs and sets of paragraphs comprising documents (such as the *Constitution*) remain matters of great debate, even when attempts are made to understand them within their historical context. It would therefore be a mistake to imagine that appeals to "context" would be more likely to support the computational conception than to undermine it. If these reflections are well-founded, therefore, then the computational conception does not appear to have a great deal to recommend it, for the twin theses that define that position are relatively clearly and decisively flawed.

THE REPRESENTATIONAL CONCEPTION

Hilary Putnam (1975) differentiates between psychological (or "mental") states for which the ascription of that state (or "token") does not presuppose "the existence of any individual other than the subject to whom that state is ascribed" (psychological states in *the narrow sense*) and the rest (psychological states in *the broad sense*) [Putnam (1975), p. 136]. As Fodor remarks, "Narrow psychological states are those individuated in light of the formality condition; viz. without reference to such semantic properties as truth and reference" [Fodor (1980), p. 331]. In view of the arguments that have been presented, it would not be unreasonable to wonder whether any account of the nature of language and of mind which, like computational accounts, restricts itself to psychological states in the narrow sense could possibly be adequate. In order to appreciate its appeal, therefore, it is crucial to consider precisely how such syntactical approaches can accommodate content.

Recall that computational conceptions not only require some strong relationship between the form of a token and its content but also need to explain how any specific content comes to be identified with any specific form. There appear to be at least three possible solutions to this problem, namely:

(4) *by nature*, according to which the connection between tokens of a specific form and their specific content is a function of the laws of nature;

(5) *by convention*, according to which the connection between tokens of

a specific form and their specific content is a function of the practices, customs, and traditions of a social group; and,

(6) *by habituation*, according to which the connection between tokens of a specific form and their specific content is a function of the habits, skills, and dispositions of individual token-users.

Indeed, the approach among these three that blends most harmoniously with syntactical models of the mind is the thesis that these relations are "natural".

Once mental tokens have been ascribed content by one or another of the modes indicated above, however, it becomes increasingly difficult to distinguish "computational" from "representational" conceptions. Stich, for example, has sought to separate "strong" from "weak" representational accounts:

> Unlike the STM, however, the weak RTM [Representational Theory of the Mind] insists that these syntactic objects *must have content or semantic properties*....The weak version claims only that every token mental state to which a cognitive theory applies has *some* content or *some* truth condition....The stronger version of the doctrine agrees with the weaker version in requiring all mental state tokens to have content or truth conditions. But it goes on to claim that these semantic features are correlated with the syntactic type of the token. [Stich (1983), p. 186]

Thus, (4), (5), and (6) reflect successively weaker and weaker modes of correlation between the syntactical type of a token and its semantical content. *Prima facie*, a purely syntactical conception of language and mentality would be an object of limited interest, since it would be unable to accommodate the ideas most fundamental to cognition and communication: content, information, and belief. Indeed, even Haugeland (1981), who has championed the computational conception of minds as "automated formal systems", has acknowledged the necessity for the formal tokens of those systems to possess semantic content: "Given an appropriate formal system *and interpretation*, the semantics takes care of itself" [Haugeland (1981), p. 24]. The problem that remains, therefore, is to explain how these correlations occur, without which the syntactical approach would simply fail to distinguish "interpreted" from "uninterpreted" formal systems [Hempel (1949a), (1949b)].

The Language of Thought. Fodor (1975) attempts to solve this problem at a single stroke by introducing "the language of thought" as an unlearned language that is both innate and species-specific. Among the most important theses upon which his position depends are (i) that learning a language presupposes the possession of a prior language, (ii) that learning a language cannot be a matter of acquiring dispositions, and (iii) that this innate language functions as a meta-language for learning other languages [Fodor (1975), pp. 64-79]. The key to Fodor's position is the assumption that learning a language requires learning the truth conditions for the sentences that can occur in that language: "...learning (a language) *L* involves learning that *'Px' is true if and only if x is G* for all substitution instances. But notice that learning this could be learning *P* (learning what *P* means) only for an organism that already understands *G*" [Fodor (1975), p. 80]. Having made this assumption, Fodor finds himself confronted by the unpalatable choice between the existence of an infinite hierarchy of meta-languages (for each meta-language in turn) and the existence of an unlearned language that serves as a base case.

Given exclusive commitment to truth-conditional semantics in the sense adumbrated here, Fodor suggests that learning a language is "literally a matter of making and confirming hypotheses about the truth conditions associated with its predicates" [Fodor (1975), p. 80], supporting the following claim:

> Either it is false that learning *L* is learning its truth definition, or it is false that learning a truth definition for *L* involves projecting and confirming hypotheses about the truth conditions upon the predicates of *L*, or no one learns *L* unless he already knows some language different from *L but rich enough to express the extensions of the predicates of L.* [Fodor (1975), p. 82]

Notice that it is the commitment to truth-functional semantics as an exclusive access route to the meaning of a sentence in *L* combined with the theoretical necessity to block an infinite regress of meta-languages for meta-languages that anchors Fodor's position. No one, presumably, would want to deny that learning a language involves projecting and confirming hypotheses about the predicates of *L*. The question is, does this have to be done as Fodor suggests?

On its face, Fodor's position appears to be very difficult to swallow. Its general character parallel's Plato's theory of knowledge as recollection.

which posits an Eternal Mind in which mortal minds participate before birth. Since the Eternal Mind is the repository of all knowledge and mortal minds participate in the Eternal Mind, all knowledge resides in every mortal mind before birth. Aware that not all mortal minds appear to be all-knowing, Plato contends that the trauma of birth induces forgetfulness in each of us, but that different experiences in life trigger off "recollections" of that lost knowledge. An alternative to Plato's account, of course, would be the theory that knowledge is acquired through experience, so that the very "experiences" in life that (for Plato) trigger off "recollections" could be the mechanisms through which knowledge is actually acquired, an account that is far more elegant.

In an analogous fashion, Fodor posits a universal language of thought of which every (neurologically normal) human being is a possessor. In order to learn any ordinary language, it is necessary to discover (through experience) the truth conditions for each of the predicates in that language within the language of thought. The language of thought itself, moreover, must be infinitely rich and extraordinarily complex, since it must have the capacity to encompass each new predicate as it is introduced into its previously impoverished ordinary-language successor (in response to the discovery and growth of science and technology, for example). Since not all speakers of ordinary language display similar linguistic ability, evidently they simply have not exercised equal ingenuity in discovering the truth conditions that lie undiscovered in their mental language as an as-yet-unrealized resource.

Since experience and ingenuity as well as the language of thought are necessary conditions for learning an ordinary language on Fodor's account, an alternative would be the theory that languages can be learned through experience and ingenuity, so that the very "experiences" in life that lead to projecting and confirming hypotheses about the predicates of L might occur even in the absence of a language of thought. For this to be possible, however, Fodor's argument must have at least one false premise, and, indeed, a little reflection suggests that Fodor may have begged the question in moving from the *premise* that learning P (learning what P means) can only occur for "an organism that already understands G" to the *conclusion* that learning P (learning what P means) requires learning meta-linguistic truth conditions for P. For that conclusion would follow only if there were no other way for an organism to "understand" the G-phenomenon that P happens to describe.

An alternative to Fodor's theory of learning a language, in other words, would be to take seriously that learning presupposes some sort of "understanding", without assuming that the form of understanding involved here has to be *linguistic*. As infants and children, we frequently – even typically – begin to do things (such a nipple, bounce a ball, smile a lot) without having any name or label for the habits, skills, or activities thereby performed. It should not be especially surprising, therefore, that when (initially unfamiliar) words are associated with (already familiar) things, including patterns of behavior that we happen to have displayed, it does not demand extraordinary ingenuity or vast experience for a (neurologically normal) human being to learn forms of language that are appropriate to their age and past experience. The problem for Fodor is the same as the problem for Plato.

The Substitutional Criterion. A possible objection might be raised that criticisms of computational accounts based on ambiguity do not affect the language of thought, simply because thought is supposed to be *unambiguous*. Even if this claim were correct, however, its truth would depend on the nature of thought as a specific account unpacks it. Fodor himself suggests that the language of thought is "very similar" to natural languages in the relevant respects [Fodor (1975), p. 156]. This justifies the use of sentences from an ordinary language like English to exemplify the difficulties that confront his position. It also hints that, if there *is* some unambiguous level of understanding, it is not likely to resemble what Fodor has in mind.

In fact, the same conclusion receives further support from a very different starting point. One of the virtues of computational and representational accounts would appear to be their capacity to exploit "the inferential network" model of language and mentality. As J. Christopher Maloney remarks,

> Computationalism considers the mind to be an inferential system. That clearly presupposes the existence of structures over which the inferences are defined....[and] the only things that seem physically fit to function as elements in material inferences are sentences. [Maloney (1988), p. 56]

Indeed, the meaning of a word or a sentence can be related to its place within a network of syntactical and of semantical relations, a paradigm of which is the standard conception of a *definition* as an entity consisting of two parts, a word, phrase, or expression to be defined (the "definiendum") and the word, phrase, or expression by means of which it is defined (its "definiens"), where both the definiendum and the definiens are envisioned as *linguistic* entities.

The catch, of course, is that it is impossible for every word in a language to be defined on pain of either an infinite regress or definitional circularity, as everyone would acknowledge. Yet every defined term can be replaced in principle by some sequence of undefined terms with which it is synonymous. Thus, by the substitutional criterion, different mental tokens have the same content when they are mutually derivable by the exchange of definiens for definiendum. Even representational accounts, however, afford no solution to the problems of ambiguity and context that we have considered above. And the meaning of undefined (or "primitive") words within a language is not settled merely by fixing their location within an "inferential network".

Assuming that words like "fast", for example, are defined terms within the corresponding language, sentences like (d) and (e) or (d) and (f) could well be mutually derivable in harmony with the substitutional criterion and presumably would possess the same content in accord with that standard. But even sentences like (f) remain ambiguous, since precisely "where" John "wants to go" might have social, sexual, or professional connotations, among others, depending upon context. The relations that obtain between the different elements that collectively comprise an inferential network are not enough to fix their content. Even if its location within a network is regarded as providing a "partial specification of meaning" for a term [Hempel (1952)], unless the meaning of some of that network's other elements is antecedently understood, such a procedure accomplishes no more than fix its syntax.

The realization that this is the case, moreover, turns out to be the basic reason why purely syntactic accounts of semantics cannot possibly succeed. Rapaport, for example, offers a sustained defense of the thesis that the only kind of semantic interpretation required for understanding natural language "is a kind that only involves syntactic symbol manipulation of precisely the sort of which computers are capable" [Rapaport (1988), p. 81]. But the arguments developed above indicate that purely syntactical manipulations can do no more than establish functions between one possibly meaningless set of marks and another. The problem that remains, therefore, is accounting for the meaning of the primitive terms that occur in a language. By positing a "language of thought", of course, Fodor attempts to finesse this difficulty, which would resurrect itself *within the language of thought* were it not for the thesis that the meaning of these tokens has been fixed "by nature"!

The persisting attraction of the inferential network conception, furthermore, may stem from the tendency to confuse *dictionaries* and *semantics* as though they were the same thing. A dictionary, after all, provides us with our most familiar example of a network of relations between words, where one entry can lead to another entry. Assuming that the dictionary is a static and fixed entity with only a finite number of entries, it will necessarily exemplify the phenomenon of definitional circularity. That is nothing new, of course; but this observation ought to be combined with the recognition that those words themselves assume significance only because at least some of them are antecedently understood independently of their location within that network. The crucial problem remains of accounting for the meaning of the primitives of the language, dictionary, or network.

The question sounds funny on its face; yet I want to contend that this is the issue that poses the deepest concern for both computational and representational conceptions. The answer, after all, apparently consists of those habits, skills, and practices by virtue of which the words we use are related to the world around us. We seldom think about defining words like "wood", "hammer", and "nail", because we can use them – and can know how to use them – without the intervention of other linguistic forms. With respect to the primitive terms that occur in any ordinary language, we must discover how they are used rather than ask for their linguistic meaning, precisely as Wittgenstein proposed. From this point of view, therefore, knowledge of a language is far more adequately envisioned as a skill than as a state (as a matter of "knowing how" rather than of "knowing that"), which itself may be the fundamental misconception lying at the foundation of these accounts.

A theory of language and mentality that could handle only uninterpreted formal systems, no doubt, would completely fail to satisfy the most elementary desiderata for theories of that kind. Yet it is important to notice a difference between at least two ways in which content could accompany a syntactical conception "by nature". A theory of the first type might assert the existence of laws of nature relating the sentential tokens that happen to occur in ordinary languages, such as English, German, and French, to mental states with specific content. The very existence of ordinary languages with such very different grammars and vocabularies, however, suggests that an approach of this kind cannot possibly be correct. There do not appear to be any grounds for accepting any semantic theory with these general features.

A theory of the second type, however, might assert the existence of laws of nature relating mental states to specific content, while making connections between those mental states and the sentential tokens that occur in ordinary languages a matter of convention. An account of this kind, in effect, would posit a set of *linguistic primitives* for an ordinary language (consisting of the undefined words in its vocabulary) and a set of *psychological primitives* for a corresponding mental language (consisting of the tokens that are related to their content "by nature") [cf. Fodor (1975), p. 124]. Obviously, Fodor's own account is of this general kind. A theory of yet a third type, however, might deny the existence of laws of nature of either kind, asserting instead that all of these connections are established either by conventions or else by habits. Only a theory of this kind offers the promise of resolving the difficulties that have been discussed above. But such an account, which emphasizes the role of individuals and of communities in establishing the significance of signs (as their users), could not appropriately qualify as a representational conception.

THE DISPOSITIONAL CONCEPTION

If these arguments are well-founded, then the dispositional conception appears to be right-headed on the whole, insofar as it provides an account that is completely compatible with the identification of the meaning of the primitives of a language with the linguistic habits that effect these connections between language and the world. Fodor's defense undoubtedly would be a thesis that we have merely mentioned in passing. For Fodor contends that learning a language cannot be a matter of acquiring dispositions: "If anything is clear", he maintains, "it is that understanding a word (predicate, sentence, language) isn't a matter of how one behaves or how one is disposed to behave" [Fodor (1975), p. 63]. And I freely admit that some of the classic positions on the nature of dispositions, such as those of Gilbert Ryle (1949), B. F. Skinner (1957), and W. V. Quine (1960), cannot cope with his criticism.

These accounts suffer from a variety of maladies, which arise because the conceptions they present are behavioristic, reductionistic, and extensional in kind. The nature of dispositions as they are understood here, by comparison, is non-behavioristic, non-reductionistic, and non-extensional [Fetzer (1981), (1986)]. Nevertheless, the adequacy of the analysis that follows can be appraised in relation to Fodor's principal argument against dispositions:

> Behavior, and behavioral dispositions, are determined by the interactions of a variety of psychological variables (what one believes, what one wants, what one remembers, what one is attending to, etc.). Hence, in general, any behavior whatever is compatible with understanding, or failing to understand, any predicate whatever. Pay me enough and I will stand on my head if you say "Chair". But I know what "is a chair" means all the same. [Fodor (1975), p. 63]

No dispositional conception of language and mentality ought to be adopted unless it can explain what is right and what is wrong with Fodor's position.

All of this would be so much "smoke and mirrors", however, were it impossible to demonstrate the benefits that accrue from adopting a pragmatic point of view. Perhaps its most crucial features were recognized by Peirce, who accentuated the place of beliefs as causal elements affecting behavior:

> Our beliefs guide our desires and shape our actions.... The feeling of believing is a more or less sure indication of there being established in our nature some habit which will determine our actions.... Belief does not make us act at once, but puts us into such a condition that we shall behave in some certain way, when the occasion arises. [Hartshorne and Weiss (1960), 5.369]

An analysis of this kind clearly falls within the broad tradition of functional conceptions of meaning and of mind, where the difference between this account and others of this general type is the specific role assigned to dispositions in understanding the nature of acts, in general, and of speech acts, in particular. While dispositional conceptions are functional accounts, in other words, not all functional accounts are dispositional conceptions – nor, indeed, do other dispositional accounts possess the special characteristics of this one.

To aid recollection, Peirce defined a "sign" as a something that stands for something (else) in some respect or other, where there are some very basic differences in the ways in which something can "stand for" something (else). Thus, he distinguished between three different classes or varieties of signs:

(7) *icons*, which are signs that stand for other things by virtue of a relation of resemblance between those signs and that for which they stand;

(8) *indices*, which are signs that stand for other things by virtue of being either causes or effects of those things for which they stand; and,
(9) *symbols*, which are signs that stand for other things by virtue of conventional agreements or habitual relations between those signs and that for which they stand.

What is most important to bear in mind within the present discussion, moreover, is that the words that are the elements of ordinary languages, such as "dog" and "chair", are symbols in Peirce's technical sense. As such, they neither resemble nor are causes or effects of those things for which they stand.

One foundation for disputing the thesis that thought requires language, therefore, derives from the realization that icons and indices are signs that do not possess the characteristics possessed by words in ordinary language by virtue of their status as symbols. So long as at least some thinking takes place in *images*, for example, where images as icons stand for that for which they stand by virtue of relations of resemblance, the computational conception cannot be altogether true. This observation, moreover, does not depend upon denying that icons and indices have syntactical properties, especially given the potential (exploited by artists, on the one hand, and by film directors, on the other) to arrange various features in various ways to create the semblance of different faces and to combine effects to simulate their causes.

Fodor, of course, has suggested that there are formal processes that are not syntactic. His position is that formal processes might be defined over representations that do not have a syntax "in any obvious sense" [Fodor (1980), pp. 309-310]. When syntax is interpreted narrowly, so that languages, in particular, have a syntax because they possess grammars, Fodor's thesis sounds quite plausible. When syntax is interpreted broadly, so that icons and indices, in particular, have a syntax because they can be combined in different ways to produce different effects, however, its loses its truthful appearance. Combining noses of various sizes, eyes of various shapes, and the like can generate enormously varied countenances. Combining sights and sounds of various kinds in various sequences can produce immensely different scenarios. There is no good reason to deny the existence of iconic or of indexical syntax or to deny the existence of iconic or of indexical form.

The basic reason for disputing the thesis that thought requires language (in the sense that ordinary languages are languages), therefore, is that their vocabularies stand for that for which they stand because of conventional or habitual connections between those words and those things for which they stand; that is, they exemplify only *one* among at least three different types of ways that something can stand for something else. This, we know, reflects the intriguing possibility that there might be more than one type of mentality corresponding to more than one type of system of signs, where semiotic systems of Type I utilize icons, of Type II utilize icons and indices, and of Type III utilize icons, indices, and symbols, as Chapter 2 explained. Indeed, we have already discovered that there may be grades of mentality that are higher than that of semiotic systems of Type III [cf. Popper (1978), (1982)].

A Theory of Cognition. The theory of signs provides a foundation for the theory of belief, however, only if a "theory of cognition" ties them together. Thus, while signs provide modes of reference for objects and for properties, they lack the assertive character of beliefs, sentences, and propositions, i.e., they are neither true nor false. The connection between signs and beliefs for a semiotic system, therefore, appears to be a causal process that arises when a system becomes conscious of a sign in relation to its other *internal states*, including its pre-existing motives and beliefs. When such a system becomes conscious of something that functions as a sign for that system, its cognitive significance results from causal interaction between that sign and these internal states, which constitute its *context*. From this point of view, therefore, the content (or meaning) of a mental state (or token) cannot be fixed independently of consideration of the context provided by a semiotic system, apart from which even its constituent signs possess no significance.

For a semiotic system to be conscious in the sense intended does not imply that it could articulate everything of which it is conscious at the time. A sign-using system is *conscious* (with respect to signs of a certain kind) when it has both the ability to utilize signs of that kind and the capability to exercise that ability, where the presence of signs of that kind within the appropriate causal proximity would lead – invariably or probabilistically – to the occurrence of cognition. When a person happens to be asleep, drugged, or intoxicated, for example, their level of consciousness may differ from what it would be were they not in those specific states. And the crucial indicator that their level of consciousness varies from case to case is that the various types of signs to which they are able to respond differs in these conditions.

By invoking this notion, therefore, I do not preclude the significance of "preconscious" or of "unconcious" states in the sense of Freudian theories, although they require interpretation in light of the conception of mentality as semiotic capability. The conception of *behavior* required by this conception, therefore, must be broad enough to encompass mental effects among its various manifestations, which occur, for example, when someone "understands something", "changes their mind", and the like. Indeed, this approach even supports the thesis that the mind can influence the body, which might take place as a result of listening to a lecture, seeing a movie, and the like, when we acquire data, information, or knowledge that we did not possess before. These are issues we shall have opportunity to consider again in Chapter 9.

Since the term "concept" would be useful to refer to the meaning of any mental token (whether sentence, sign, or belief), let us adopt it here. Thus, a complete account of *the content of a concept* for a specific semiotic system (if it were possible) would be provided by an inventory of all of the kinds of behavior toward which that system would be disposed under all of the different kinds of contexts within which it might find itself. When the content of a concept would be displayed in various different ways under infinitely varied conditions, then any merely finite description of its significance could never be more than partial and incomplete. The most suitable approach toward understanding the content of a concept is in relation to its causal role:

> (Thus,) the most perfect account of a concept that words can convey will be a description of the habit which that concept is calculated to produce. But how otherwise can a habit be described than by a description of the kind of action to which it gives rise, with a specification of the conditions and of the motive? [Hartshorne and Weiss (1960), 5.491]

Indeed, it seems clear that, in the case of human semiotic systems, the range of behavioral manifestations that a concept has for a system would have to vary across the (complete sets of) motives, beliefs, ethics, abilities, capabilities, and opportunities that influence its behavior as a complex causal system.

From this perspective, the theory of meaning presupposes the theory of action. The theory of action that shall be adopted here takes it for granted that *human behavior* and *human actions* are not identifiable, insofar as the class of human actions is restricted to the class of human behaviors that are brought about – possibly probabilistically – by the causal interaction of one's own motives, beliefs, ethics, abilities, and capabilities, where the success (or failure) of those efforts tends to depend upon and vary with the opportunities with which we are confronted (including, in particular, with whether or not the world is as we believe it to be, i.e., as a function of their truth). Ultimately, of course, it may be important to distinguish more precisely between these assorted types of factors, but let us assume they form a complete set.

To illustrate the character of the account which I am endorsing, observe that a marksman who wants to hit his target, who believes that his target is present, and who does not rule out firing at this target on moral grounds can hit his target only when his skills are equal to the task, his rifle and ammunition are available, and the target itself is within his vicinity [Fetzer (1986), p. 106]. When individuals happen to be neurologically impaired, physically restrained, morally debauched, deliberately misinformed, etc., then the kinds of behavior that they tend to display under otherwise similar conditions varies from those that tend to be displayed by individuals who are not neurologically impaired, physically restrained, morally debauched, etc. The behavior someone displays on a specific occasion thus results from the complete set of relevant factors present on that occasion, where a factor is relevant if its presence or its absence on that occasion made a difference to the strength of the tendency for that person to display behavior of that kind.

The Functional Role Criterion. Thus, the content (or the meaning) of a specific sign (or token) can be captured by identifying its causal role in influencing different kinds of behavior under different kinds of conditions, where those conditions (for human systems) tend to be complex. A few examples may serve to illustrate the conception I am recommending here:

Example (C): John has the misfortune to be confronted by a burglar in his own apartment. The burglar, whom John does not recognize, menaces him with a knife, so he escapes by climbing out the window, which is open.

Example (D): Mary has the misfortune to be confronted by a burglar in her own apartment. The burglar, whom Mary does not recognize, menaces her with a knife, but she has a broken leg and cannot climb out the window.

In appraising whether or not the same sign (the burglar with the knife) has the same content (or meaning) for John and for Mary, a comparison has to be drawn, not between their *actual behavior* (which was obviously different) but between their *dispositions toward behavior* (how they would have behaved had they had the chance, relative to the same motives, beliefs, etc.), which requires intensional (subjunctive and counterfactual) rather than extensional (historical and indicative) formulations. If John and Mary would have behaved in the same ways across every complete set of relevant conditions when conscious of that sign (when their contexts were the same for both), then that sign would possess the same content (or meaning) for both – even though the behavior they actually displayed may have been different! In particular, if Mary would have escaped out the open window if she had not had a broken leg, if John would have done whatever Mary actually did had he found himself in her situation (including her abilities and capabilities), etc., then that sign would have had the same content for both of them.

What this implies is that Fodor is mistaken in thinking that there could be no behavioral constancy that would provide a foundation for a dispositional account. Fodor's mistake results from adopting an excessively behavioristic and extensional account of dispositions. "Pay me enough and I will stand on my head if you say, 'Chair'", says he. "But I know what 'is a chair' means all the same." Exactly! So too for everyone else whose motives and morals would inspire them to similar behavior – knowing what "is a chair" means just as well! It also suggests the solution to a difficulty observed by Donald Davidson long ago, according to which the meaning of a token (sign) cannot be a disposition to use specific words in specific ways, *simpliciter*, if only because not everyone speaks English! Of course, he is right, except that this dispositional conception only requires that different speakers use similar words on similar occasions in similar contexts – *provided they have the same linguistic abilities*! Otherwise, no such similarity needs to follow.

Hence, if we want to isolate special kinds of causal factors, such as the content (or meaning) of specific beliefs $B1$, $B2$, ..., then we can do so by holding constant those other beliefs Bm, Bn, ..., motives $M1$, $M2$, ..., ethics $E1$, $E2$, ..., abilities $A1$, $A2$, ..., capabilities $C1$, $C2$, ..., and opportunities $O1$, $O2$, ..., whose presence or absence make a difference to the (internal or external) behavior that would be displayed by that system, given the presence of Bi (where i ranges over 1, 2, ...). Then the content (or meaning) of a specific belief, $B2$, say, is the totality of tendencies that the system would possess in the presence of that belief, and the difference that having that belief rather than some other, say, $B1$, is the difference between the totality of tendencies that that system would possess in the presence of *B*1 and the totality of tendencies that that system would possess in the presence of *B*2!

Thus, for any specific semiotic system, a sign, *S*, stands for something *x* for that system rather than standing for something else *y* if and only if the strength of the tendencies for that system to manifest behavior of some specific kind when conscious of *S* – no matter whether publicly displayed or not – differs in at least one context; otherwise, there is no difference beween *x* and *y* for that system. When two signs, *S*1 and *S*2, possess exactly the same meaning (as two tokens of exactly the same type), then the strength of the tendencies for that system to manifest behavior of various kinds when conscious of *S*1 must be the same in every context as is the strength of its same tendencies when conscious of *S*2. This account therefore allows – indeed, it requires – that "sameness of meaning" be amenable to *degrees of similarity* that occur when two tokens are tokens of some of the same types but not of all. When comparisons of (what might be called) "cognitive significance" in its *narrow sense* (encompassing differences in meaning other than purely linguistic ones, involving the specific words used, their precise sounds, etc.) are desired, that measure can be obtained, in principle, by subtracting all these purely linguistic behavioral phenomena from (what might be called) "cognitive significance" in its *broad sense* (which encompasses all meaning). Indeed, these conceptions afford a foundation for resolving issues of ambiguity and of synonymy across the board within a dispositional framework.

Methodologically, at last, this conception exemplifies what Carl G. Hempel describes as, "the epistemic interdependence of belief and goal attributions" [Hempel (1962), Sec. 3.3] – or, better, the epistemic interdependence of motive, belief, ethics, ability, capability and opportunity ascriptions. This means that, in order to subject an hypothesis about causal factors of any of these kinds to empirical test, it is necessary to make assumptions concerning the simultaneous values of each of the others. Fortunately, this result, which is not theoretically avoidable, is not theoretically objectionable. But if such a conception is even roughly right-headed in its approach, it provides a rather striking explanation for the inherent complexity of social and psychological phenomena. From this point of view, the complexity of the phenomena with which social science must contend in comparison to that with which natural science must contend becomes strikingly apparent – a remarkable outcome!

FUNDAMENTALLY DIFFERENT SYSTEMS

During the course of this investigation, we have undertaken an exploration of the solution space for the problem of discovering an adequate conception of the nature of language and mentality. Attention has been given to approaches of three distinct types, computational, representational, and dispositional conceptions, which accent syntactical, semantical, and pragmatical aspects of the phenomena with which they are concerned, respectively. We have discovered that the computational conception, even when complemented by the parsing criterion, adopts an extremely implausible theory of the relation of form to content, which cannot contend with problems of ambiguity, with problems of synonymy, or with problems of context. We have also discovered that the representational conception, which benefits from an "inferential network" approach and from the substitutional criterion, cannot resolve the underlying difficulty of fixing the meaning of primitive language.

These reflections have led to an exploration of the dispositional conception which, unlike its alternatives, appears to succeed where they have failed. In particular, by appreciating the role of habits, skills, and dispositions in establishing connections between language and the world, the pragmatical approach has the capacity to deploy the functional role criterion in dispatching problems of ambiguity and of synonymy, while taking proper account of the place of context. It overcomes the *prima facie* case against dispositional conceptions by embracing an account of dispositions that is at once non-behavioristic, non-reductionistic and non-extensional, which enables it to overcome not only Fodor's complaints but Davidson's objection as well. Thus, an analysis of this kind, which unpacks the content (or meaning) of mental tokens by means of their causal role within an intensional framework, is not vulnerable to the criticisms that undermine different accounts, including Quine (1975).

One of the most important consequences that attend this investigation, moreover, emanates from the crucial role of individual sign-users as semiotic systems. For the notion of an *idiolect* as the sign system utilized by a single sign-user turns out to be more fundamental theoretically than is that of a *dialect* (understood as a regional phenomenon) or that of a *language* (if understood as a social group phenomenon). Thus, there is nothing here that inhibits the prospects for the existence of "private languages", in the sense of constellations of dispositions for speech and other behavior that reflect surface manifestations of possibly unique correlations of primitives to the world. Indeed, the results that have been uncovered here clearly suggest that convention has a secondary role to play by contrast with habituation, promoting communication by providing a social mechanism for resolving differences and for codifying practices concerning what does and does not qualify as "standard usage" within particular language-using communities.

Another important consequence attends the realization that physical systems can be distinguished as systems of (shall we say) "fundamentally different" kinds when the specific types of factors that make a difference to the behavior of those systems varies from case to case. Human beings, from this perspective, qualify as motive-belief-ethics-ability-capability and-opportunity-types of systems, since these reflect the range of causal factors that affect the behavior of systems of this kind. Digital computers, by comparison, seem to be hardware-firmware-software-and-input/output-device-types of systems instead, since they reflect the range of causal factors that affect the behavior of systems of that kind. The proper measure of similarity and difference, once again, of course, has to be subjunctive and counterfactual rather than historical and descriptive in relation to complete sets of relevant conditions that influence systems of these kinds.

Thus, digital machines would be "fundamentally similar" to human beings only if they were subject to the same range of factors, which is plainly not the case. Indeed, this discovery appears to provide a plausible explanation for the deeply held intuition that human beings and digital machines really are "fundamentally different" as types of knowledge, information, and data-processing causal systems. But this notion itself appears to harbor an equivocation. In its *weak* sense, these systems are "fundamentally different" merely because they are implemented by means of different material (or "media"). In its *strong* sense, these systems are "fundamentally different", not merely because they are implemented by means of different material (or "media"), but because their modes of operation – their processes and programs – are not either similar or the very same.

This ambiguity arises because the specific types of factors that make a difference to the behavior of systems of these different kinds include their modes of operation. Otherwise, of course, any specification of the range of causal factors that affect their behavior would be seriously incomplete. That human beings and digital machines are "fundamentally different" in the weak sense thus undermines the prospect of emulation, but that human beings and digital machines are "fundamentally different" in the strong sense undermines the prospect for replication as well. No one wants to deny that digital machines specifically and computational devices generally are capable of simulating the cognitive processes of human beings. If human beings are pragmatic semiotic systems, while computational devices are syntactic symbol systems instead, however, then they are incapable of replicating one another's modes of operation.

The results of this chapter are meant to imply that systems of these kinds are "fundamentally different" in both – strong and weak – senses. Whereas most students of AI tend to believe that AI concerns how we do think rather than how we should think or even how we could think, the results of this investigation suggest another possibility. If, after all, human beings and digital machines are "fundamentally different" in the strong sense indicated here, what grounds remain in support of the view that digital machines and human beings *can* process knowledge, information, or data in similar ways? The great debate between the "strong" and the "weak" conceptions of AI, it appears, may rest upon a misconception, which would indeed be the case if systems of one kind (digital machines) are incapable of functioning in the same way as systems of the other kind (human beings). If this were the case, it might not only be false that AI concerns how we do think but also false that it concerns how we should.

Could AI then concern how we might think? The answer to that question, I believe, depends upon whether or not the architecture of the brain is compatible with these possible modes of operation of the mind. As far as the dispositional conception of language and mentality is concerned, it seems quite plausible to imagine that a connectionist parallel-processing architecture might do better than the classical sequential digital machine architecture that tends to be presupposed by computational conceptions. Connectionism models the brain as a neural network of numerous nodes that are capable of activation. These nodes can be connected with other nodes, where, depending on their levels of activation, they may bring about increases or decreases in the levels of activation of the other nodes [cf., esp., Rumelhart et al. (1986), Smolensky (1988), and Fetzer (1990)].

These patterns of activation, in turn, might function as signs for those larger systems of which they are otherwise meaningless elements by coming to stand for various things for those systems of which they are parts. These might include features of their internal states or of their external environments as a consequence of the ways that these things can fuction as signs for those systems. When viewed from this perspective, an architecture appropriate for mental systems viewed as semiotic would require the ability of relating patterns of one kind to patterns of another, the ability for such patterns to affect the behavior of the systems, and the ability for these patterns to stand for other-than-pattern things by functioning as signs for systems of that kind within the context of a pragmatic conception. These desiderata strongly suggest the possibility that differences between connectionist models and digial machines may be among those that differentiate mental systems from other kinds of causal systems [Fetzer (1990)].

The demise of the computational conception and the rise of connectionist conceptions may be promising for resolving the problem of replication. Mere lack of success in pursuing this objective, moreover, can never prove that this goal cannot be accomplished, though it can diminish our prospects. But the pursuit of machine mentality does not define the scope and limits of AI. The proper conclusion to draw is that AI should also develop machines that can solve problems in ways that are not accessible to human beings. There are ample grounds to justify the belief that even digital devices are undeniably things that can do many things successfully on behalf of human beings. The loss of an illusion that cannot be sustained (if that, indeed, is what it turns out to be – and we can never be certain) does not demolish the immense potential of this important discipine. Perhaps the appropriate attitude to adopt is to subordinate the pursuit of creating thinking things and get on with the business of designing intelligent machines.

PART II

KNOWLEDGE AND EXPERTISE

4. THE NATURE OF KNOWLEDGE

One of the principal motives for pursuing an analytical problem-space approach for this investigation is that some of the most influential figures working within this field have gradually evolved in their positions, Fodor especially. Fodor (1987), for example, yields a functional analysis within common-sense psychology, yet he persists in maintaining that "there has to be a language of thought" [Fodor (1987), pp. 135-154]. On this point, I believe he is completely mistaken. The only presupposition for learning a language appears to be species-specific *predispositions* [cf. Fetzer (1985b)]. But it remains essential to understand the results discovered thus far. In Chapter 2, the symbol system conception was shown to be inferior to the semiotic system conception as an account of the nature of mind. Much of of Chapter 3, by comparison, was devoted to establishing the inadequacy of any computational approach to the analysis of language and mentality. Both were directed against certain conceptions of AI rather than AI itself.

If the analysis of Part I withstands critical examination, anyone who adheres to either the "strong" or the "weak" thesis is likely to be discouraged. But it is important to notice that the results we have derived strongly suggest that the symbol system conception – or one very similar – *is* an appropriate account of the mode of operation of digital machines. The computational approach appears to succeed as an analysis of machine intelligence as opposed to that of human minds. The scope of AI needs to be defined in relation to the limitations inherent to mechanisms of that kind. If minds are semiotic systems and if language and mentality are species of complex dispositions, then *both* the strong thesis and the weak thesis stand in need of reconstruction. Indeed, assuming our findings are well-founded, digital machines are incapable, in principle, of replicating the mental activity of human beings. They cannot replicate how humans do think, and they cannot replicate how humans should think. Restricted to the architecture of devices of this kind, the AI quest for replication appears to be in vain.

This discovery, however, ought to be liberating rather than stultifying. For these digital machines can successfully accomplish any number of important and valuable tasks – not, indeed, in the mode of replication, but in the mode of simulation. Indeed, these potential applications do not begin to exhaust what these machines can do in exceeding the performance capabilities displayed by human beings. The purpose of Part II, therefore, is to elaborate some of the specific types of roles for which computers may properly be utilized. These include the rapid processing of vast quantities of data and the simulation and modeling of natural and of social phenomena, because of which computers are destined to become as significant as telescopes and cyclotrons within empirical sciences. But they also encompass a diversified range of other sorts of activities, especially those within the scope of "expert systems", broadly defined. And understanding these systems requires familiarity with the nature of knowledge representation.

There are few problems within computer science and artificial intelligence that raise as many troublesome and fascinating issues as does that of the representation of knowledge. While books and entire issues of journals have been devoted to this subject, its theoretical foundations remain largely unsettled and matters of dispute. Although there appears to be widespread agreement on the four most important modes of knowledge representation,

(I) (1) semantic networks,
 (2) scripts and frames,
 (3) production systems, and
 (4) predicate calculus,

as Gordon McCalla and Nick Cercone have observed [McCalla and Cercone (1983), p. 13], there is nothing approaching a consensus concerning when one of these modes is to be preferred to another, much less a theoretical framework with respect to which their relative strengths and weaknesses might be systematically assessed. [See especially *COMPUTER* (October 1983), Brachman and Levesque (1985), Levesque (1986b), and Cercone and McCalla (1987).]

Actually, this state of affairs does not seem so surprising when consideration is given to the unusual set of issues that are confronted in this particular domain. For these matters involve difficulties that are ordinarily investigated within disciplines as diverse as metaphysics and epistemology, the philosophy of language, and logic and logical theory. They are by no means restricted to computer science and artificial intelligence. Indeed, as we are going to discover, which distinctions need to be drawn with respect to the representation of knowledge depends upon the purpose that is intended to be achieved. The representation of ordinary knowledge, for example, may be different from that of scientific knowledge. It might prove to be useful, nevertheless, to discover what might serve as an *idealized framework* for representing different kinds of knowledge, where those confronted with a specific problem within this domain might then adopt precisely the distinctions that matter, from their specific point of view. The purpose of the discussion that follows, therefore, is to explore the relations obtaining between the theory of knowledge and the theory of knowledge representation and to examine the connections between the philosophy of language and the representation of knowledge, with special concern for problems in representing scientific knowledge. The result is intended to be a systematic framework for pursuing these issues from the perspective of the theory of knowledge.

CONDITIONS OF KNOWLEDGE

The theory of knowledge may be regarded as an attempt to ascertain the different means by which the kinds of knowledge that humans can acquire might be discovered against a background of assumptions about what there is to discover, as a function of a theory about the world's general structure. This intimate relationship between *epistemology* (as a theory of the means whereby knowledge might be discovered) and *ontology* (as a theory about the world's general structure) suggests that the pursuit of epistemology in the absence of ontology is not very likely to succeed. And there appears to be a parallel relationship between the theory of knowledge representation and the theory of knowledge that implies similar results when a theory of knowledge representation is pursued apart from the theory of knowledge.

	Theory of Knowledge:	*Theory of Knowledge Representation*:
Objective:	To discover the ways in which things can be known	To discover the ways in which knowledge can be represented
Assumption:	That there are various things that can be known	That there are kinds of knowledge to represent

Fig. 9. Dependency Relations I.

From this point of view, the theory of knowledge hinges upon logically prior commitments to ontological assumptions about the broad categories of the kinds of things (including objects and events) that define the world's structure and the relations that may obtain between them (including logical relations and causal connections) that arise within the universe as a whole or within various domains of discourse about aspects of what there is. The relations involved here are those described by Fig. 9. This is an account to which we shall have occasion to return in Chapter 5. [Cf. esp. Scheffler (1965) and Dancy (1985) on the nature of epistemology.]

The Classic Definition. The most influential construction of the nature of knowledge is the classic account, which entails the conception of *knowledge* as warranted, true belief. In other words, a person *z* is supposed to know *p*, where '*p*' asserts that something is the case, when and only when:

(II) (i) *z* believes that *p*,
(ii) *p* is the case, and
(iii) *z* is justified in (i).

The various modes of warrant whereby *z* might be justified in believing that *p* include perceptual inference and inductive and deductive reasoning. The key feature differentiating perceptual inference from inductive and deductive reasoning, moreover, is that perceptual inference involves a (more or less direct) causal relationship between a perceiver and the world, while induction and deduction involve ways of reasoning from premises instead.

For example, if '*p*' is a description of something (such as the color of an apple) that *z* could be justified in believing merely by looking in a certain place at a certain time (say, in the kitchen where the apple could be found) under suitable conditions (when the lights were on, *z* was not blindfolded, etc.), then that belief could be justified by a perceptual warrant. However, if '*p*' is a general claim (say, about the color of every apple of a certain kind, such as Golden Delicious) that *z* could be justified in believing if *z* were in possession of appropriate information (say, that many different persons at many different times had examined apples of this kind under suitable conditions, when they were not selected for color, dyed a certain shade, etc.), then *z* might be entitled to reason beyond what he or anyone else had ever directly observed to the conclusion that all such apples were of that color.

Even with respect to the strongest kind of warrant that any belief would seem to be able to receive, the standard conception of knowledge reflected by (II) has been subject to dispute. Bertrand Russell, for example, offered an obvious illustration of a difficulty with that account, which takes as its setting the British political scene of 1912:

> If a man believes that the (then) late Prime Minister's last name began with *B*, he believes what is true, since the (then) late Prime Minister was Sir Henry Campbell Bannerman. But if he believes that Mr. Balfour was the (then) late Prime Minister, he will still believe that the late Prime Minister's name began with a *B*, yet this belief, though true, would not be thought to constitute knowledge. [Russell (1959), pp.131-132]

In cases of this kind, of course, there is a deductive warrant between what *z* believes to be the case (regarding the late Prime Minister) and something else that *z* infers by deduction from that belief, where *z*'s conclusion follows from its premises. It thus qualifies as a warranted true belief that appears, nevertheless, not to be knowledge. Examples of this kind have preoccupied epistemologists since an article by Edmund Gettier (1963), and entire books have been devoted to "the Gettier problem", namely: does warranted, true belief always qualify as "knowledge"? [Cf., for example, Shope (1983).] The solution required seems to depend upon the character of warrants. We will eventually consider its ramifications for program verification in Chapter 8.

Apart from this debate, however, the conception reflected by (II) is not difficult to defend, since it is relatively easy to display the inadequacies of alternative conceptions, at least with respect to ordinary knowledge. For if "knowledge" merely required belief, then even those whose beliefs are contradictory would be the possessors of knowledge, but that would mean that the very same thing – such as the color of that apple – could be known to be both red and yellow (or any other color anyone believes it to have). And although that effect would not be problematic for the notion of belief, it would generate intolerable consequences with respect to knowledge itself. Similarly, true beliefs unsupported by warrants are unlikely candidates for knowledge, not least of all because the absence of evidence would make it impossible to tell which beliefs among them all are properly labeled "knowledge".

Certainty vs. Fallibility. The role of warrants is that of epistemic criteria. The classic conception of knowledge permits a distinction to be drawn between beliefs that are warranted (and true) and beliefs that are warranted (and false). The beliefs we accept are not always true. Our criteria are not presumed to be *infallible*, even though the Cartesian attitude toward the nature of knowledge denies that knowledge can ever be uncertain. Since perceptual inference and inductive reasoning are incapable of supporting inferences to conclusions that cannot be false, the adoption of that position has the unenviable consequence of rendering most of what we think we know to be merely a matter of belief. Descartes contended that the only beliefs of which he could be *certain* were those he was incapable of doubting, including, as it turned out, the belief that he existed as a doubting thing. For the more he doubted whether he doubted, the more convinced he became. With help from some suppressed premises, moreover, Descartes was able to reason his way to "certain belief" in the existence of God and the world.

A similar position was advanced much earlier by Plato, who maintained (in addition to his theory of recollection) that *knowledge* could be attained only with respect to an eternal and unchanging "World of Being". The world of perceptual experience, as opposed to that of inner thought, can only gain access to the transient and fleeting "World of Becoming", about which only uncertain *opinion* could be maintained. And this remains the case, even if the evidence upon which those beliefs are founded is entirely appropriate. One of the most important differences between Plato and Descartes in this regard, however, is that Plato's position depends upon the properties of the different things in the world that might be known, whereas Descartes' rests upon differences in attitude on the part of the thing itself that might know. Plato's emphasis, therefore, is on *the things known* (as objective properties of the world) rather than *the knowing thing* (and its subjective properties).

No doubt, we confront a difficulty whose solution ultimately requires a stipulation as to how the word "knowledge" is going to be used. But unless we are prepared to deny that we can ever know the color of an apple that may be found in our kitchen or that every Golden Delicious has a specific color, the positions embraced by Plato and by Descartes should not be embraced. What is most important about the classic definition from this perspective is the latitude that it permits for warrants that are less than conclusive for beliefs that may still be true, where "uncertain knowledge" is not a self-contradictory expression. Beliefs whose warrants are less than conclusive, therefore, will include those based upon perceptual and inductive findings, which will still qualify as "knowledge" only if they are true. [The truth condition receives further discussion in the following chapter.]

The insuperable problem before us is that, without direct access to the true, we have no grounds to distinguish warranted beliefs that are false from warranted beliefs that are true. Since warranted beliefs are beliefs that we ought to accept on the basis of the available evidence, we can be rationally obligated to embrace at least some false beliefs. This result reflects one of the most important reasons why freedom of information is so fundamental to the democratic way of life. Without free exchange of information among citizens, scientists, and politicians, those who control access to information control the evidential basis for rational belief. But it also confirms what we have discovered already, since it suggests that the capacity to make mistakes is an essential aspect of the human condition.

A Refined Conception. The notion of a warrant can be made far more precise, even in its general character. Consider, for example, the following four conditions of "rational belief" specified in relation to a knower z at a time t in relation to a set of sentences S and a possible new belief b by now assuming that all of our beliefs can be represented by sentences:

(III) (CR-1) the set of sentences S accepted by z at t is deductively closed,
(CR-2) the set of sentences S accepted by z at t is logically consistent,
(CR-3) if some of the evidence available to z at t warrants the acceptance of b, then z must accept b into S, so long as
(CR-4) all of the evidence available to z at t also warrants the acceptance of b into S at t,

where these conditions ought to be understood as applying in relation to a presupposed language framework and justifiable rules of inductive and of deductive inference [Fetzer (1981), pp. 9-16]. And it is not very hard to establish that each of these requirements fulfills an indispensable role.

Condition (CR-1), for example, which is the closure condition, requires that any consequence of any belief that we accept must also be accepted. In accepting a belief, of course, we are accepting it as true, so this condition simply insists that everything that must be true when that belief is true is part of what we accept when we accept it. Condition (CR-2), the consistency condition, requires that we never accept sets of beliefs that cannot possibly be altogether true. If we were to violate this condition, we would thereby accept some beliefs as well as their negations, where their negations assert that those beliefs are false. Jointly these requirements serve to guarantee that the sets of beliefs that we accept are maximally informative and possibly true, which are both desirable properties.

Condition (CR-3), the partial evidence condition, entitles us to accept beliefs for which we have enough evidence (relative to the language and inference rules that we employ) so long as condition (CR-4), the complete evidence condition, is also satisfied. Condition (CR-3) is important in permitting the acceptance of a belief when there is enough (experiential or sentential) evidence to support its truth without insisting that we must first inventory all of the evidence available to us. When all the evidence available to us would invalidate that inference because it has been superseded by another, then (CR-4) requires its revision. This condition, which conveys the same force as Rudolf Carnap's "requirement of total evidence", is especially important in perceptual inference and in inductive reasoning.

The role of (CR-3) and (CR-4) can be illustrated quite readily in relation to perceptual (or "observational") inference. In the case of perception, as we gather more and more relevant data, our conclusions are subjected to revision. Suppose we are walking along the street and notice someone in the distance who looks like a very close friend. At some point still fairly far removed, we might well conclude that it is our friend approaching us. Yet upon even closer inspection we might belatedly realize to our chagrin that this was not our friend at all. What finally gave the lie to our earlier conclusion, moreover, may be something very small, such as a gesture or a mannerism others might not notice at all. And this case exemplifies the Principle of Minimality mentioned in Chapter 2, namely: that the smallest detail can make the greatest difference in matters of perceptual inference. We shall refer to the results of these inferences as "experiential findings".

It ought to be observed, however, that none of these conditions has the effect of imposing the requirement of truth. In this sense, they jointly define the conception of rational belief *without* entailing that rational beliefs have to be true beliefs. Such an additional condition could be specified by,

(CR-5) the beliefs accepted by z at t must be true.

Yet, without direct access to the truth, we have no independent method of determining which of those beliefs that satisfy (CR-1) through (CR-4) also satisfy (CR-5). Since those that do satisfy (CR-1) though (CR-4) will be believed to be true, of course, it will then be reasonable to *believe* that (CR-5) has been satisfied as well. But nothing involved here provides a guarantee. In view of the possibility of acquiring additional evidence over time, however, the interplay of these conditions will have the effect of requiring us to accept, to reject, or to suspend beliefs in response to available evidence.

This conception thus implies what shall be called "a theory of epistemic resources" concerning the materials that are available in the pursuit of rational belief (or "knowledge"). Our principal resources appear to be these:

(IV) (1) the semiotic abilities A that a person z can exercise at t,
(2) the deductive rules of inference upon which z relies at t,
(3) the inductive rules of inference upon which z relies at t,
(4) the experiential findings available to z at t, relative to A, and
(5) the powers of imagination and conjecture z can exercise at t,

where the role of imagination and conjecture in conjuring up original ideas and novel hypotheses suggests another capacity machines may not have. In elucidating this conception, however, it might be worthwhile to define a few features that distinguish these different types of rules of inference.

Both inductive and deductive reasoning proceed from given premises (in the form of sentences) to derived conclusions (in the form of sentences), where acceptable rules of inference correspond to acceptable forms of argument. The features distinguishing (good) deductive forms of argument are:

(a) they are *demonstrative*, i.e., if their premises were true, their conclusions could not be false;
(b) they are *non-ampliative*, i.e., there is no information or content in their conclusions that was not already contained in their premises; and,
(c) they are *additive*, i.e., the addition of further information in the form of additional premises can neither strengthen nor weaken these arguments, which are already maximally strong.

Thus, the non-ampliative property of (good) deductive arguments can serve to explain both their demonstrative and additive characteristics. Demonstrative arguments, of course, are said to be "valid". And valid arguments with true premises are said to be "sound". The important property of sound arguments, in turn, moreover, is that their conclusions cannot possibly be false.

By contrast with deductive arguments, (good) inductive arguments are:

(a′) *non-demonstrative*, i.e., their conclusions can be false even if their premises are true;
(b′) *ampliative*, i.e., there is some information or content in their conclusions not already contained – implicitly or explicitly – in their premises; and,
(c′) *non-additive*, i.e., the addition of further information in the form of additional premises can either strengthen or weaken these arguments.

Arguments satisfying appropriate inductive standards likewise may be said to be "proper". And proper arguments with true premises may be said to be "correct". But a correct inductive argument can still have a false conclusion.

It should come as no surprise that the purposes served by such different types of argument are quite distinct, indeed. For inductive arguments are meant to be *knowledge-expanding*, while deductive arguments are meant to be *truth-preserving*. This description of the function of induction, it should be observed, would be self-contradictory if knowledge could never be false. The ways in which inductive arguments expand our knowledge assume various forms. Reasoning from samples to populations, from the observed to the unobserved, and from the past to the future always involves drawing inferences to conclusions that contain more information or content than do their premises. The most familiar instances of inductive reasoning that all of us employ in our daily lives, moreover, concern the behavior of ordinary physical things – things that may or may not work right, fit properly, or function smoothly (including products of technology, like microwave ovens and personal computers). When we interact with systems such as these, we invariably base our expectations upon our experience: we draw conclusions concerning the future based upon their behavior in the past. All such reasoning is ampliative and – as we often discover to our dismay – is not demonstrative.

In spite of the familiarity of induction in our daily lives, the principles in terms of which it should best be understood are far from common knowledge. Indeed, their precise character is a subject of enormous debate among theoreticians and philosophers. Rather than engage in this dispute and enter into the complexities that make it such a difficult area of inquiry,

I shall simply illustrate a few inductive policies (IP) that theoreticians might consider:

(IP-1) *the straight rule*: inductive inference is a matter of extrapolating from the relative frequencies that have obtained in the past to the relative frequencies that will obtain in the future by assuming that they will remain essentially unchanged;

(IP-2) *Bayesian inference*: inductive inference is a matter of utilizing Bayes' Theorem as a formal property of the calculus of probability, where beliefs about the future based upon beliefs about the past can be adjusted by a process of conditionalization; and,

(IP-3) *abductive inference*: inductive inference is a matter of utilizing the principle of maximum likelihood in order to formalize a pattern of reasoning known as "inference to the best explanation", whose mathematical properties are unlike those of probabilities.

These are only three of the possible alternatives, where orthodox statistical hypothesis testing represents yet another inductive policy. Indeed, there are various competing objective and subjective versions of Bayesian inference itself. [Cf. Fetzer (1981), Part III. Vaughan (1988) considers the implementation of Bayesian and likelihood policies with computer programs.]

The general conception that emerges is a framework for understanding modes of argumentation for reasoning from sentential premises to sentential conclusions – where unsuccessful arguments are classified as *fallacious* because of problems of irrelevance, ambiguity, etc. – which looks like this:

From SENTENTIAL PREMISES
to SENTENTIAL CONCLUSIONS

DEDUCTIVE:	*INDUCTIVE:*	*FALLACIOUS:*
a) demonstrative	*a) non-demonstrative*	*a) are neither*
b) non-ampliative	*b) ampliative*	*b) irrelevant*
c) additive	*c) non-additive*	*c) ambiguous*

straight rule	*Bayesian inference*	*abductive inference*
(extrapolative)	*(probabilistic)*	*(likelihoods)*

Fig. 10. Modes of Argumentation

The principal mode of *non*-argumentative inference, therefore, is what we have been referring to above as "perceptual inference". When the results of perception are expressed by means of sentences, these experiential findings in turn can serve as evidence sentences in inductive and in deductive arguments. In order to pursue the theory of epistemic resources, however, a distinction must be drawn between *languages* and *language frameworks*, which differ from languages in several significant – even crucial – respects. Much of the debate over the nature of language arises from their confusion.

LANGUAGE FRAMEWORKS

The importance of language (and of language frameworks) to understanding the nature of knowledge and of knowledge representation cannot be over-emphasized, especially because beliefs are normally expressed by means of a language employed by a person z at a time t. As we have already discovered in Part I, there are reasons for supposing that the conception of a system of signs is more general than the standard conception of a language. Indeed, the use that shall be made of the phrase "language framework" is somewhat different from that which has been made of the term "language". A natural (or "ordinary") *language*, strictly speaking, is a special kind of dispositional ability possessed by semiotic systems of Type III. In the discussion that is to follow, however, we shall be concerned with a syntactical conception that may be viewed as a counterpart to the computational conception, but understood to be inadequate for completely understanding the nature of language. Since digital machines can fulfill the computational but not the dispositional conception, we need to explore precisely how far that resource can carry us.

A *language framework* **L** may be entertained as *an abstract model* of the syntax of a language that is of special interest for our purposes. When the crucial differences between languages and language frameworks need to be considered, we shall return to this distinction. Given this understanding, a language framework **L** (or a "language **L**", for short) has these dimensions:

(V) A language **L** includes:
  (1) a vocabulary of
    (a) logical terms, and
    (b) non-logical terms; and

(2) a grammar of
  (a) formation rules, and
  (b) transformation rules;

where the formation rules determine which strings of marks from the logical and non-logical vocabulary qualify as well-formed formulae (or "sentences") of **L**, and the transformation rules determine what follows from what. Since the tranformation rules may be viewed as deductive rules of inference that might be distinguished from the rest of language **L**, let us regard a *language framework proper* as consisting of elements (1)(a) and (b) with (2)(a) alone.

This account is therefore a purely syntactical conception, which we know is inadequate to capture semantical and pragmatical features of a language, as Chapter 3 has explained. As we have already discovered, the non-logical terms of a language consist of primitive and defined terms, where not all of the terms in any language can be defined within that language without generating either a definitional circularity or an infinite regress. The meaning of primitive terms for human languages, of course, tends to be secured by means of habits, skills, and dispositions. A language **L**, by comparison, does not attempt to capture the semantic and pragmatic dimensions of ordinary language, but instead may function as a special type of *artificial language*, whose various features may or may not be interpreted as standing for the corresponding features of some other language. Although a language **L** in this sense reflects at most merely the syntactical skeleton of any ordinary language, if it reflects any ordinary language at all, it can be useful, nevertheless, especially as an appropriate framework for artificial intelligence. [The best description of this relationship remains that of Carnap (1939).]

Analytic Sentences. This background sets the stage for a discussion of two of the most important and influential distinctions in the history of philosophy between "*a priori*" and "*a posteriori*" kinds of knowledge, on the one hand, and "analytic" and "synthetic" types of sentences, on the other. The distinction between analytic and synthetic knowledge is generally attributed to Immanuel Kant, who drew the distinction in relation to sentences of subject-predicate form. On Kant's conception, if the predicate-concept has already been thought in the subject-concept, a judgment is analytic; otherwise, it is synthetic [Walsh (1967), p. 309]. Some examples might include, say, "All freshmen are students", "All bachelors are unmarried", and such. The subjective and psychological character of Kant's conception, however, has motivated the search for a more objective and linguistic construction.

A somewhat different formulation of this notion that promises to fulfill this desideratum would be that sentences are analytic when they are themselves logical truths or are reducible to logical truths by the substitution of definiens for definiendum, whereas synthetic sentences concern the world:

(D6) analytic sentences =df sentences that are either logical truths or reducible to logical truths by substitution of definiens for definiendum relative to a language; and,

(D7) synthetic sentences =df sentences that provide information or content about the world whose truth does not follow from the adoption of that language alone.

The distinction between analytic and synthetic sentences, therefore, is relative to a presupposed language (or language framework **L**) and requires an acceptable conception of *logical truth* for its implementation. The notion of logical schemata whose substitution instances cannot be false is sufficiently familiar from elementary logic to serve as an illustration. If we employ an arrow "→" as the material conditional sign (where "... → ___ " stands for "if ... then ___" when it is interpreted as a *material conditional,* which must be true if either its antecedent is false or its consequent is true), then it seems reasonable to assume that, say, "If snow is white, then snow is white" is an example of a sentence that cannot possibly be false (so long as the meaning of the words therein, such as "snow" and "white", is constant throughout, a condition that is known as "the requirement of a uniform interpretation").

The adequacy of this approach has been challenged by W. V. O. Quine in one of the most influential papers of 20th C. philosophy [Quine (1953)]. He begins by considering a preliminary conception of analyticity, namely: that a sentence is analytic when it is true by virtue of meanings and independent of fact. Finding the notion of meaning to be ambiguous, he resorts to a familiar distinction between a term's *extension* (the class of all entities to which it applies) and its *intension* (the set of conditions on the basis of which it applies to any given entity), for which he retains the term "meaning". In this fashion, he differentiates between the theory of reference and the theory of meaning, asserting the primary business of the latter to be "the synonymy of linguistic forms and the analyticity of statements" [Quine (1953), p. 22].

Tackling first the notion of analyticity, Quine discovers two classes of allegedly analytic sentences, namely: those that are logically true and those that are reducible to logical truths by means of the substitution of synonyms for synonyms. (Observe that since the definiens of a definition has the same meaning as its definiendum, the conception advanced above may be viewed as a specific instance of this more general characterization.) He offers as an example of the first kind the sentence, "No unmarried man is married", and as an example of the second kind the sentence, "No bachelor is married". He contends that, whereas in the case of logical truths a clear criterion of analyticity can be specified (since a logical truth is a sentence that is true and remains true under all consistent reinterpretations of its components that are not logical particles, themselves defined by enumeration), no similar method exists for generating the other class of sentences that does not itself rely upon a notion of synonymy "which is no less in need of clarification than analyticity itself". Hence, the existence of a clear distinction is placed in jeopardy.

To support this charge, Quine argues especially that synonymy cannot be analyzed adequately by reference to definition, interchangeability, or semantic rules, because (i) with the exception of explicitly conventional notation introduced for the sake of convenience as in the case of science (which he calls "an extreme sort of definition"), definitions simply report on or improve upon linguistic usage and where synonymy is concerned implicitly rely upon a notion of synonymy-in-use; (ii) synonymy when analyzed by means of interchangeability *salva veritate* (while preserving truth value) presupposes the notion of analyticity (since the truth of the new sentence has to be the same as that of the old sentence, insofar as these substituted parts have the same meaning as those they replace); and (iii) semantic rules, even within artificial languages, merely indicate which sentences are analytic without elucidating the concept itself (typically, by providing a list of the sentences of that kind).

Thus, in Quine's view, we are confronted with a dilemma: on the one horn, we find it impossible to explain analyticity without making reference to synonymy; on the other, we are unable to explain synonymy in the absence of reference to analyticity. He therefore concludes that belief in the distinction between analytic and synthetic sentences is a mere dogma, "a metaphysical article of faith", that ought to be abandoned [Quine (1953), p. 37]. In spite of the enormous influence that this article has exerted, there are at least four reasons to doubt that Quine's conclusion ought to be embraced:

First, even if we were to grant his premises, his conclusion would not follow. Quine's critical remarks are directed only against analyticity in a broad sense rather than in the narrow sense represented by logical truths. As a consequence, his conclusion can at most refer to this broader category rather than to all analytic sentences.

Second, by unpacking the notion of logical truth, Quine himself draws a clear boundary between one class of analytic sentences as distinct from the class of synthetic sentences, one that does not appear to be "metaphysical" at all.

Third, by admitting the conventional introduction of convenient notations as in the case of science, Quine overtly permits an important class of analytic sentences that are plainly analytic in the broad sense and not simply logical truths.

Fourth, his argument appears to be inconsistent to the extent to which he displays willingness to unpack the concept of a logical particle by enumeration thereof but finds the same procedure unacceptable if applied to the concepts of synonymy and of analyticity in the broad sense.

In spite of this, Quine's position represents a deeper difficulty in dealing with concepts such as these. The methodological commitments that reflect the foundation of these arguments are only implicit in Quine's arguments here but become explicit in his other work, including Quine (1960). For the methods to which he is committed are extensional, behavioristic and reductionistic in character. [For some responses to Quine's position, see Harris and Severens (1970); for more recent views, see, for example, Schwartz (1977).]

Since we have already discovered the limitations inherent in the parsing criterion and the substitutional criterion in matters of this kind, it ought to come as no surprise that methods such as Quine's are inadequate to resolve the problems to which they are being applied. An adequate account of the nature of synonymy and analyticity presupposes adoption of the functional role criterion in their place, which, in turn, depends upon the utilization of (counterfactual and subjunctive) dispositional conceptions that are at once non-extensional, non-behavioristic, and non-reductionistic. It should not be surprising that these notions would appear to be somewhat vague and amorphous in the absence of a methodology that would be adequate to cope with them. Indeed, as Chapter 3 suggests, these notions can only be unpacked by appreciating more fully the extent to which the theory of meaning depends upon the theory of action and on the adoption of an adequate methodology.

Nonetheless, these reflections afford an opportunity to emphasize an important consequence attending the dispositional conception of language and mentality. For, speaking generally, except in those cases when the signs under consideration are two syntactical tokens of the same syntactical type, it will almost always be the case, even when those tokens happen to have the very same cognitive significance *in the narrow sense* (encompassing differences in meaning other than purely linguistic ones, involving specific words, their precise sounds, etc.), that they will tend to differ in their cognitive significance *in the broad sense* (precisely because this sense includes all of the purely linguistic differences excluded by the narrow sense). Even when the word "bachelor" has the same meaning as the expression "unmarried adult male", it does not mean that those words and expressions are used under all and only the same conditions, since these will include differences in context that involve pragmatic dimensions of communication situations and the like. When there is more than one way to say the same thing, we tend to use the words that best achieve our goals in the context in which we find ourselves.

What this implies, therefore, is that cognitive significance, from the dispositional perspective represented by the functional role criterion, clearly can be a matter of degree, where degrees of similarity and of difference in cognitive significance are a fundamental fact of life. When the definiens of a definition contains more words or more letters than its definiendum, it must be expected that the sounds made and the words said under various conditions of behavior and speech will not be the same. From the deeper perspective afforded by a dispositional account, it becomes clear that this does not mean that those words or those expressions differ in their meaning. It only means that the distinction between cognitive significance in its broad sense has to be isolated from cognitive significance in its narrow sense, precisely as we have done.

Even more important than the realization that the analytic/synthetic distinction can be shown to be justified for ordinary languages, however, is the realization that its applicability follows as a trivial consequence from the adoption of a language framework **L**. In defining the features of a language of this kind, we are using symbols to stand for whatever we may choose. Some of these symbols may stand for elements of its grammar and others for elements of its vocabulary. Both the parsing and the substitutional criteria can fulfill their intended roles. Moreover, were we to construct such a language and implement its rules by means of a program, the result would be the creation of an automated formal system. A system of symbols of this kind, no doubt, would satisfy Newell and Simon's conception. Yet these signs would only be significant for the users of that system and not for the system itself. An artificial language of this kind can fulfill our objectives in relation to AI without incurring the types of controversy that arise for ordinary language.

A Priori Knowledge. For languages of both of these kinds, the distinction between analytic and synthetic sentence is important for at least two different reasons. The first concerns the kind of content (or "information") that knowledge of these distinct types can provide. The second concerns the way in which that content can be acquired (or "learned"). In particular, a distinction must be drawn between *a priori* and *a posteriori* as types of knowledge:

(D8) *a priori* knowledge =df knowledge whose justification is independent of experience in the form of observations and experiments; and,

(D9) *a posteriori* knowledge =df knowledge whose justification is dependent upon experience in the form of observations and experiments.

The most obvious candidates for *a priori* knowledge are the sentences whose truth follows from adoption of a language (or language framework L) alone, i.e., those that are logical truths or are reducible to logical truths by the substitution of definiens for definiendum, which qualify as "true by definition".

Apart from these special classes of analytical sentences, it is difficult to discern other clearly justifiable cases of *a priori* knowledge. The classic philosophical position known as "empiricism", moreover, maintains that analytic sentences can afford the only form of *a priori* knowledge. The position that is known as "rationalism", by contrast, asserts the opposite. Their positions on these matters thus reflect a stance toward the existence of analytic and synthetic sentences as representing different kinds of content and *a priori* and *a posteriori* knowledge as reflecting different conditions of acquisition. Contemporary interest in these positions has been sparked by debate over the nature of language itself, where those who maintain that possession of language is innate (including Fodor) appear to adopt the rationalist position, whereas the dispositional conception is an empiricist view [cf. Stich (1975)].

This issue is somewhat cloudy because "*a priori* knowledge" is usually taken to include all of the consequences that follow from the adoption of a vocabulary and grammar alone without presuming that those sentences are informative about the world. From this perspective, the difference between viewing language as innate or as learned seems to involve whether this is something we are born with as a form of analytic *a priori* or *a posteriori* knowledge! For the innateness hypothesis to postulate any form of synthetic *a priori* knowledge requires the additional assumption that there is the right sort of fit "by nature" between the signs that appear in the innate language and the things that occur in the world, just as Fodor claims. For reasons set forth in Chapter 3, Fodor appears to be on the wrong track in maintaining this position. But it ought to be emphasized that whether it qualifies as synthetic *a priori* knowledge rests upon special assumptions.

The position that a theoretician adopts on matters such as these, nevertheless, can be reflected by the intersection of these distinctions. Thus, the empiricist conception maintains that any knowledge that can be acquired independently of experience must be analytic, but any knowledge that is synthetic must be directly or indirectly related to observation and experiment:

	Analytic Sentences:	*Synthetic Sentences:*
A priori Knowledge:	Yes	No
A posteriori Knowledge:	No	Yes

Fig. 11. The Empiricist Position.

The phrase "empirical knowledge" thus refers to knowledge of the truth of synthetic sentences, acquired on the basis of observation and experiments. The rationalist position, by comparison, can be characterized in this fashion:

	Analytic Sentences:	*Synthetic Sentences:*
A priori Knowledge:	Yes	Yes
A posteriori Knowledge:	No	Yes

Fig. 12. The Rationalist Position.

The existence of defensible examples of synthetic *a priori* knowledge today remains a matter of debate, a problem we shall not continue to pursue here. [For related discussion, however, see Fetzer (1990).]

Modes of Modality. At least equally important to an investigation of relations between the theory of knowledge and the theory of knowledge representation, moreover, are certain modal distinctions that apply within the category of empirical knowledge itself. For things of a specified kind *K* can have the properties that they have in at least three different ways, namely:

(VI) (a) as a matter of definition,
(b) as a matter of coincidence, or
(c) as a matter of natural law.

Whether something (or some class of things) has a certain property *P* (at a certain time *t*) as a matter of definition, as a matter of coincidence, or as a matter of natural law, however, depends in part upon the language (or the framework **L**) that is presupposed and in part upon how that thing (or class of things) happens to be referred to or described. To simplify the presentation of this information, I shall do so in relation to a framework **L**. But the same points could be made with reference to an ordinary language instead.

If the term "chair" were to mean "a raised surface suitable for sitting by one person, which is easily moveable" within a certain framework, **L**1, let us say, it would then be the case that if something were a chair (in relation to that framework), it would have to have the properties of being a raised surface, of being suitable for sitting by one person, etc., merely as a matter of definition. That this would be the case is simply a consequence of adopting that definition within that framework. A thing of the kind *chair* could fail to have those properties, relative to **L**1, only if that framework were either abandoned or revised, but then it would be a different framework. It need not be the case that to be a chair something would have to be made out of wood, covered with black naugahyde, etc., but it would have to be easy to move, etc., as analytical consequences attending the use of that framework.

Notice further, moreover, that if some specific thing *x* were to be identified by name as "Mother's chair", for example, then it would also have to have whatever other properties that specific name was intended to entail. If that thing *x* happened to be a chair that was given to someone known as "Mother", then for something to be that thing it would have to be not only a raised surface that is suitable for sitting, etc., but also the one that was given to the person referred to by that name. Indeed, speaking generally, we can say that the function of *proper names* is to provide a means to refer to some specific thing to the exclusion of other things: these linguistic entities can be most successfully understood as "definite descriptions" in the sense of Russell (1905). On this account, the meaning of a proper name is the same as a description beginning with the definite article "the" (for example, in the case at hand, "Mother's chair" means "the chair that was given to Mother"), implying thereby that there is at least and at most one such thing of that specific kind. [A summary of his position may be found in Russell (1919), Chapter 16. Important discussions of his account are found in Klemke (1970) and in Linsky (1977). A variety of alternative positions are presented in Schwartz (1977).]

In order to determine whether or not a specific thing x remains the same specific thing even though some of its properties may change across time (as one of its legs might be broken and have to be replaced), the kind of thing of which that specific thing is a particular instance must be made explicit. This can be done by formally specifying the conditions that underlie its existence as a thing of a specific kind. Otherwise, since any specific thing ceases to exist as a thing of a certain kind K when it no longer satisfies the corresponding description, it would be theoretically impossible to ascertain whether or not the thing named by a specific proper name continues to exist as a three-legged thing, as a wooden thing, or whatever, because of various changes in the properties that it may have at different times [cf. Fetzer (1981), pp. 43-44]. In this particular case, therefore, so long as that thing endured as a thing of the kind *chair* and was the one and only thing of that kind given to the person known as "Mother", it would still continue to exist as Mother's chair.

Things can have properties as matters of natural law in addition to the ones they possess as matters of definition. For example, if "Elmer" were introduced into a framework **L2** as a proper name for John's pet duck, let us say, then Elmer must not only possess the properties that it has as a duck within **L2** but also whatever other properties things that happen to be of that kind happen to have as a matter of natural law. If "duck" means, say, "a small swimming bird with a flat bill, short neck and legs, and webbed feet" in **L2**, for example, then those properties of being a small swimming bird, of having a flat bill, and so forth, are properties that Elmer must have as a matter of definition. But there are other properties that things of this kind could not lack as a matter of natural law, such as needing air, water,and food to survive; mates of the opposite sex to reproduce; and the like. These are properties that Elmer could not be without while remaining an instance of a duck (thus defined), yet they are not properties that Elmer has by definition – unless things of that kind were so defined within **L**2.

The properties that something has as a matter of natural law, let us note, are *relative properties*: whether they are properties of a thing or not depends upon the kind of thing of which that thing is supposed to be an instance. And this, in turn, depends upon the meaning assigned to the words that occur within that framework. The introduction of the name "Elmer" as the name of a specific individual thing of the kind *duck* (understood as it has been previously defined), therefore, means (i) that Elmer, as a duck, will have certain properties that he cannot be without as matters of definition; (ii) that Elmer, as a duck, will have certain other properties that he cannot be without as matters of natural law; and (iii) that Elmer, as a duck, will also have a variety of other properties that could be viewed as coincidental (or even as "accidental"), relative to the framework *L*2 and the laws of the world. Thus, Elmer is sometimes sitting up and sometimes sitting down; he is sometimes eating and sometimes sleeping; he sometimes weighs a few ounces more and sometimes a little less.

His posture, his activities, and his weight are all properties that Elmer may have at one time and not at another without jeopardizing his existence as a thing of the kind *duck* or as a distinctive instance of that kind, namely: as John's pet duck. If Elmer's ownership were to change, naturally, then he would no longer exist as John's pet duck, although his existence as a thing of the kind *duck* would remain unchanged. Let us therefore say that properties that things of a certain kind have as a matter of natural law are "permanent properties", while those that things of a certain kind might gain or lose without losing membership in that kind are "transient properties". Then properties of both of these kinds can only be specified in relation to some presupposed *reference property*, since properties of both of these kinds are relative properties. [The ontology that underlies this distinction is discussed in Fetzer (1981), pp. 36-45.]

Thus, to offer an additional illustration, suppose that things of kind *gold* are things whose molecular structure is that of atoms with atomic number 79. Then it would be logically impossible for anything that is gold to fail to have atoms with that atomic number, as a matter of definition. And it would be nomologically impossible for anything that is gold to fail to have the melting point of 1063 C, the specific gravity of 19.32, and the like, as a matter of natural law. For the only way in which properties such as these can be taken away from things of that kind is by bombarding their nuclei or otherwise altering their atomic structure, i.e., by making them no longer *gold*. Yet there are plenty of other properties that things of this kind might gain or might lose without ceasing to be things of that kind, including their size, their shape, and their current price per weight. For there is nothing about things that are *gold* that requires them to be smaller than a donut, shaped like a ring, or sell for $500 an ounce, which they all could be without.

If every thing of a specific kind *K* happened to have a certain property in common, then it might be difficult to discover whether or not those things had those properties by definition, by natural law, or by accident. Suppose, for example, that the kind *Volkswagen* were defined as automobiles meeting certain specifications with respect to their engine design, chassis construction, and the like, but without reference to color. Then, if all Volkswagens happened to be grey, for example, it might look as though this color were a permanent property of things of that kind, until consideration is given to the existence of processes or procedures by means of which that property could be taken away from things of that kind without making them no longer Volkswagens.

The realization that repainting them would be one such process or procedure, therefore, would justify the inference that color, unlike gas economy and handling characteristics, is merely a transient property of things of this kind, whether or not it were ever to come to pass that any of them is actually repainted. Because this is the case, the general claim that all Volkswagens are grey could have been a true generalization about the world's history, yet it nevertheless would not have been a law. It is therefore essential to distinguish between these two kinds of generalizations on the one hand and definitional truths on the other.

Generalizations that are true by definition cannot be false without changing our framework, but they are uninformative about the world. Nomological and accidental generalizations, by contrast, are both informative about the world when they are true, yet the information which they provide is of very different kinds. For nomological generalizations reflect permanent property relations as a matter of natural law, whereas accidental generalizations merely reflect transient property relations. So three types of generalization should be distinguished in relation to **L**:

(D10) definitional generalizations =df generalizations whose truth is a consequence of the adoption of that language framework;

(D11) nomological generalizations =df generalizations that are not true by definition but whose truth reflects the laws of nature; and

(D12) accidental generalizations =df generalizations that are not true by definition or by virtue of natural laws but by coincidence.

Unlike definitional and nomological generalizations, accidental generalizations continue to be open to violation and subject to change, even when the language framework **L** and the laws of nature remain the same. With generalizations of this kind, it is not too misleading to suggest that even if they have been true in the past, they do not have to be true in the future.

THE CONDITION OF TRUTH

One way to account for this puzzling phenomenon is by reference to the distinction between "occasion sentences" and "eternal sentences", which has been drawn by Quine (1960). A sentence is *eternal* if and only if its truth value remains the same from speaker to speaker and from time to time. A sentence whose truth value could vary, even while remaining within the very same language framework, by comparison, is merely an *occasion* sentence. Familiar examples are first-person-pronoun sentences: when I say,"I am hungry", its meaning is not the same as when you use the same words, and the same specific sentence may be true at one time and not at another, even when said by the same speaker. Accidental generalizations are like this in the sense that, even if the world's history might have made it true that, say, "All Volkswagens are grey" circa 1962, that would not mean the truth of this claim could not change with time and be false now circa 1990.

There are several ways to draw the line between analytic and synthetic sentences, on the one hand, and nomological and accidental generalizations, on the other. Both kinds of synthetic generalization are *logically contingent* in the sense in which analytic generalizations are *logically necessary*. Accidental generalizations, moreover, are *physically contingent* in the sense in which nomological ones are *physically necessary*. The differences involved here can likewise be drawn in terms of "possible worlds" of different kinds. Sentences that are logically contingent are then true in some (logically) possible worlds but false in others, while logically necessary ones are true in all (logically) possible worlds. Sentences that are nomologically contingent are then true in some (physically) possible worlds but false in others, while nomologically necessary sentences are true in all (physically) possible worlds.

These relations can be represented by a diagram of the following form, where "L-determinate" and "L-indeterminate" are here implicitly defined:

<table>
<tr><th colspan="2">THE FALSE</th><th colspan="2">THE TRUE</th></tr>
<tr><td>necessarily
false

false in all
L-possible
worlds

L-determinate</td><td colspan="2">logically contingent
synthetic

true in some logically
possible worlds,
false in some logically
possible worlds

L-indeterminate</td><td>necessarily
true

true in all
L-possible
worlds

L-determinate</td></tr>
<tr><td>A PRIORI</td><td colspan="2">A POSTERIORI</td><td>A PRIORI</td></tr>
<tr><td>narrow L-false   broadly L-false</td><td colspan="2">physical impos-sibilities   physical contin-gencies   physical neces-sities</td><td>narrow L-true   broadly L-true</td></tr>
<tr><td>ANALYTIC</td><td colspan="2">CONTINGENT</td><td>ANALYTIC</td></tr>
</table>

Fig. 13. Modes of Modality.

Thus, this diagram reflects most, but not all, of the distinctions between the analytic and the synthetic, the *a priori* and the *a posteriori*, and the necessary and the contingent that we have considered, [For discussion of the variety of senses these terms have been given, see Sumner and Woods (1969).]

Truth and Meaning. Although considerable attention has been focused upon the nature of belief and the nature of warrants, the nature of truth as a condition for the possession of knowledge requires elucidation. As in the case of these other notions, the purpose of this discussion is to provide elements that might serve as a foundation in the theory of knowledge for considering issues in the theory of knowledge representation. The place to begin the discussion of truth within this context is the theory advanced by Alfred Tarski (1956), known as "the semantic conception". Tarski championed the distinction between *object* and *meta-languages*, where meta-languages are used to talk about object languages and the difference is a relative one. We shall then turn attention to the work of Davidson on truth and meaning.

In his very famous paper, Tarski suggests that the truth conditions for a sentence formulated in a formalized language **L** can be specified relative to a "counterpart" sentence (let us say) in another language **ML**. The language **ML** must be essentially stronger than the language **L**. This means that **ML** must have the capacity to express everything that **L** can express – lest **ML** be unable to formulate a counterpart for every sentence of **L** – and it must have the additional apparatus required to talk about the truth values of sentences in another language. Tarski's schema thus connects sentences *S*1, *S*2, ... in **L** with sentences *p*1, *p*2, ... in **ML**, where they are related so as to satisfy the following condition (where *i* ranges over the numerals 1, 2, ...):

(VII) sentence *Si* is true-in-**L** if and only if *pi*.

Hence, the sentence "Snow is white" is true in English, for example, if and only if snow is white, when "enhanced English", as it might be called, is our meta-language. This relationship thus reflects a specific instance of Tarski's schema:

(VIII) "Snow is white" is true in English if and only if snow is white,

where the sentence in quotes belongs to the object language (which in this case is English) but the entire sentence belongs to the meta-language itself.

It has been claimed (with justification) that, at the very least, Tarski established that "truth" is a meta-linguistic predicate. This means that it is sentences – as linguistic entities – that are the bearers of truth (more specifically, that sentences-as-they-occur-in-a-language-**L** are the sorts of things that are true or false). Since specifying the truth conditions for a sentence appears to be one way of assigning its meaning, Davidson (1967) has gone so far as to maintain that Tarski has in fact provided a *theory of meaning* as well as a *theory of truth*. That Davidson could be mistaken, however, follows from the realization that the Tarski schema works – to the extent to which it does work – only because of some capacity to pair up sentences in **L** with their counterparts in **ML**. This tacitly presumes that the meaning of sentences occurring in **L** and in **ML** is already known: indeed, how else would it be possible to "pair them up" the right way? Even our famous example illustrates this point. For if the word sequence "snow"/"is"/"white" did not have the same meaning in **ML** that it has in **L**, the proferred truth condition would be nothing of the sort. Tarski's conception does not solve this problem; on the contrary, it assumes that the problem has been solved.

A similar point has been made by William Lycan, who has noticed that this difficulty remains even when such a quoted sentence is properly understood to be a *structural-descriptive name* of the sentence under consideration, where a structural-descriptive name describes all of the semantically relevant features of that sentence in **L** [Lycan (1984), pp. 38-39]. Surely, if we do not know the meaning of a sentence, we cannot possibly know all of the "semantically relevant" features of that sentence, in which case no solution of the kind that Davidson suggests is in sight. To appreciate the dimensions of this difficulty, consider the following sentences from Chapter 3:

(IX) (1) That car is hot!
(2) That car has heated to over 100 degrees Fahrenheit!
(3) The cops are out looking for that car everywhere!

Presumably, the truth conditions for sentence (1) are not difficult to display:

(**ML**-1) Sentence (1) is true in English if and only if (1),

which, of course, is very plausible. But what if the meaning of sentence (1) as it occurs in **L** differs from the meaning of (1) as it occurs in *ML*? Then the first occurrence of "That"/"car"/"is"/"hot" could differ in meaning from its second occurrence. Suppose, for example, that the first occurrence has the same meaning as sentence (2) and the second the same as sentence

(3). Clearly, under these conditions it would be a complete mistake to suppose,

(**ML**-2) Sentence (2) is true in English if and only if (3).

Unless we are willing to conclude that (2) and (3) have "the same meaning", we ought to admit that Tarski's scheme works, after all, only because it presupposes that the meaning of every sentence in the object-language as well as that of possible counterparts in the meta-language is already known. For otherwise we have no grounds to suppose that they are properly rather than randomly paired. Hence, if this argument is right, then Davidson's is wrong.

Davidson could respond to criticism of this kind by proposing the following potential defense, which is intended to blunt the problem of ambiguity:

> As long as ambiguity does not affect grammatical form, and can be translated, ambiguity for ambiguity, into the metalanguage, a truth definition will not tell us any lies....let our metalanguage be English, and all *these* problems will be translated without loss or gain into the metalanguage. [Davidson (1967), p. 316]

But while Davidson may rejoice at the thought that the solution to this problem is so easily arranged, it works satisfactorily only so long as the meaning of object sentences in English is the same as that of their grammatically isomorphic counterparts in meta-English. What Davidson overlooks and what Tarski's theory does not supply is a guarantee that the creation of an essentially stronger language by supplementing that language with the apparatus necessary to formulate truth conditions themselves leaves the other elements of that language – especially, the meaning of its constituents – completely unaffected. Otherwise, the occurrence of the same words in the same sequence no more guarantees that Tarski's conception will be fulfilled than the corresponding correlation guarantees that sentences (2) and (3) are synonymous.

Another source of support for this contention comes from the domain of *machine translation*. [See, for example, Slocum (1985), Hutchins (1986), and Nirenburg (1987).] For machines to translate from a "source language" into a "target language", they must be provided with a "machine dictionary", which contains every entry in the source language, their equivalents in the target language, and a set of codes that enable the machine to process the text into the target language [Garvin (1985), p.

239]. Troublesome problems arise within this field from the generation of word-to-word translations that often prove to be grossly inadequate. As a consequence, machine translation ordinarily requires supplementation in the form of "linguistic processing", which involves the resolution of problems of ambiguity and of problems of context by "corrections" to the machine's program by human translators:

> The need for linguistic processing is more clear-cut in the case of machine translation.... To achieve better than gross word-for-word translation, each sentence of the text has to be processed linguistically to ascertain its syntactic structure *and meaning* before effecting the transfer of meaning from the "source language" to the "target language" text. [Garvin (1985), p. 239; emphasis added]

Indeed, a straightforward application of Tarskian schemata when the object language is English and the meta-language is Russian, for example, has led to unintentionally hilarious pairings when the meaning of counterparts was not already known. These include the translation of such sayings as, "The spirit is willing but the flesh is weak", into Russian sentences that translate back to English as having the meaning, "The vodka is good but the meat is rotten". [For examples, see Slocum (1985), Nirenburg (1987), or even Twain (1875).]

Austin's Conventions. Another issue that these examples tend to clarify is that truth and meaning alike seem to be properties not just of sentences as-they-occur-in-a-language-**L** but of sentences-as-they-occur-in-a-language-**L**-in-use-on-a-specific-occasion. After all, one and the same sentence, such as (1), might be used on two different occasions with very different meanings, such as (2) and (3). On both such occasions, (1) occurs-as-a-sentence-of-English, yet sometimes it occurs-as-a-sentence-of-English and has the same meaning as (2) and other times it occurs-as-a-sentence-of-English and has the same meaning as (3). While it would be ridiculous to maintain that the meaning of sentence (2) is the same as the meaning of sentence (3), it would be absurd to deny that sentence (1) occurs-as-a-sentence-of-English on both occasions. The appropriate move, under these conditions, therefore, would appear to be not only to deny the truth of Davidson's theory of meaning but also to qualify the truth of Tarski's theory of truth. These considerations, moreover, are applicable to other accounts that depend upon similar appeals to elaboration, translation, or paraphrase, for the very same reason.

The point can also be made relative to the somewhat more sophisticated analysis of truth proposed by J. L. Austin (1950). Austin analysed the use of language as a matter of "doing things with words" (as actions involving speech). He distinguished between (what he called) *locutionary acts* (the use of specific words on specific occasions), *illocutionary acts* (the contents that were conveyed by the use of those words on those occasions), and *perlocutionary acts* (what was accomplished – or intended to be accomplished – by means of the use of those words on those occasions). He was particular adept at isolating and identifying different uses of language, including, for example, to amuse, to deceive, to mislead, to entertain, to insult, and to persuade. Austin was convincing in establishing that words are used for many purposes other than simply telling the truth, but here truth is our concern.

The importance of what Austin suggests on this topic can be appreciated by realizing that, for every extensional (or truth-functional) language, such as first-order predicate calculus, the meaning of every molecular sentence is reducible, in principle, to the meaning of its own components in the form of atomic sentences, whose own meanings are left undefined. This reflects the truth-functional character of the relations between molecular and atomic sentences within extensional languages, of course, so from at least one point of view this result should not be surprising. But it means that the problem of meaning and truth for molecular sentences is directly dependent upon the meaning and truth of its atomic components, whose own meaning and truth are left unexplained. Logical relations are accounted for, but not meanings.

This situation may be acceptable with respect to the artificial languages we construct to serve our own purpose. As Lycan has remarked, when we create our own formalized languages, we can impose upon them whatever characteristics we may desire: "Clearly the (primitive) expressions of our canonical idiom will be univocal, since by hypothesis that idiom is a regimented and logically perfect formal language" [Lycan (1984), p. 36]. Yet it appears to be equally clear that the solutions that might be satisfactory in the case of artificial languages may or may not apply to corresponding problems in the case of ordinary languages. One of the merits of Austin's analysis, from this point of view, is that it applies to sentences of the simple atomic form, '*Fa*' (or '*Fat*'), where he suggests that any sentence of the form '*Fa*' is true when the object that is denoted by the name '*a*' has the property that is designated by the predicate '*F*' (at the time denoted by '*t*').

What the name '*a*', the predicate '*F*', and the indexical '*t*' stand for turns out to be a function of two types of conventions that Austin distinguishes:

(a) *descriptive conventions* correlating words (or sentences) with the various types of situations, kinds of things, sorts of events, and so forth that might possibly be encountered in the world; and
(b) *demonstrative conventions* correlating the use of those words (or sentences) with specific situations, individual things, particular events, and so on, that actually are encountered in the world.

Thus, Austin suggests that a sentence *S* is properly qualifed as true within a language when the historical situation with which *S* is correlated by its descriptive conventions is of a type with which *S* is correlated by its demonstrative conventions. This means that Austin, in effect, has reduced the nature of truth to the theories of how predicates designate and of how names denote. If names can be adequately understood to be definite descriptions, therefore, then the nature of truth depends upon the theory of designation. Although we shall not pursue this matter further here, Tarski arrives at a similar result in a more technical analysis, which focuses on the satisfaction of sentential functions by individual things. Perhaps the key difference between them, therefore, is that Tarski applies himself to a formalized language framework **L**, whereas Austin investigates ordinary language instead.

Appropriate Beliefs. It is easy to see that carrying this further tends to lead us toward a pragmatical conception, since Austin's "conventions", as properties of users of language, can readily be understood as specific linguistic dispositions, where a "language" qualifies as a special kind of skill, habit, or ability. These dispositions, no doubt, include those that relate words to the types of situations, etc., to which they are properly applied (as dispositions *to know how to use them* under certain – perhaps highly complex – types of circumstances), and those that relate the use of words to specific situations, etc., in which they can be (properly or improperly) applied (as dispositions *to use them* under certain – perhaps highly complex) circumstances. Precisely because the use of specific words on specific occasions occurs as an effect of the complex interaction of motives, beliefs, ethics, abilities, capabilities, and opportunities, however, drawing inferences about the elements of an individual language-user's language is always a difficult task. The epistemic interdependence of all the factors affecting verbal behavior means that special conditions must be assumed with respect to motives, beliefs, ethics, abilities, capabilities, and so forth, as Chapter 3 explained. Otherwise, it is impossible to justify the conclusion that the linguistic practices that are accessible to experience reflect the normal rather than abnormal, deceptive, or abusive use of language by a specific individual as a manifestation of attempts to deceive, amuse, etc.

What this means is that descriptions of languages that are based upon observations of verbal behavior depend upon or tacitly assume how people are disposed to behave in their speech and other behavior. If the members of a specific language-using community wanted to deceive an outside intruder in his attempts to understand their language, that would not be so very hard to do. The frequencies with which specific forms of speech are observed to be employed on specific types of occasions (where those "types of occasion" are defined by publicly accessible criteria) might or might not be indicative of persistent and enduring linguistic practices rooted in the dispositions of the members of that community. Without the benefit of helpful subjects (who *want* to inform, who *believe* that is a good idea, and who *know* the language well, for example), the results of an empirical investigation of the linguistic practices of a community are likely to be the flawed outcome of inadequate sampling [like the experience of Margaret Mead in Samoa; Freeman (1983)].

This phenomenon, of course, is simply one more consequence of the fact that the theory of meaning presupposes the theory of action, as we already know. Distinctions can be drawn between at least three different types of linguistic dispositions, however, since "rules of language" occur in human beings in the form of dispositions for arranging signs in sequences ("syntactical dispositions"), dispositions for relating signs to their definitions ("semantical dispositions"), and dispositions for relating signs with behavioral contexts in general ("pragmatical dispositions"). While the meaning of the primitive signs that occur within any system of signs can only be ultimately understood by means of a pragmatic account of language and mentality, "meaning" in these syntactical and semantical senses can be isolated within an individual's *idiolect* by subtracting "meaning" in its narrow sense (which encompasses differences other than purely linguistic ones) from "meaning" its broad sense (which encompasses all meaning), yielding the conception of an individual's collection of tendencies to relate those signs in various ways.

From this point of view, it looks as though it ought to be possible, under suitable conditions, to derive inferences concerning the syntactical and the semantical dispositions of individual sign-users as those notions have been defined above. These inferences, however, are obviously theoretically loaded in assuming that the data upon which they are based is sufficient in size, random in selection, and representative in character. What this clearly indicates, of course, is that reasoning of this kind has to be inductive and not deductive. It likewise should be evident that any attempt at generalization from the linguistic practices of one individual to those of a group of individuals can be a hazardous pastime. Ontologically, this is because the linguistic habits of a member of a community need not be the same as those of other members of that community. Epistemologically, it reflects the nature of induction.

The only secure means for generalizing across a population with respect to its linguistic practices is to maintain the trifling position that someone belongs to the same linguistic community as does the subject of interest if and only if they adhere to the same linguistic practices. This maneuver ensures that every member of the linguistic community thereby specified will have the same linguistic dispositions. In this case, however, the members of that language-using community will possess the same dispositions by definition. Indeed, if a distinction is drawn between *the original community* of interest (which might be specified by means of its geographical boundaries or other non-linguistic properties) and *the linguistic community* thereby defined, it should be obvious that there is no guarantee that even two members of the original community of interest will turn out to share the same dispositions.

All of this serves to emphasize the importance of conventions in setting standards for the practices, customs, and traditions of a social group. It is important to observe, however, that "conventions" can be understood in at least two quite different ways. For, in its normative sense, "conventions" represent the practices to which we ought to adhere (if we want to be, say, proper in our practice). In its descriptive sense, by contrast, "conventions" represent the practices to which we actually do adhere (which might be a mixture of "proper" and "improper" practices, from the normative point of view). The methods appropriate to discovering descriptive conventions are empirical and theoretical. Those appropriate to normative conventions, by contrast, tend to be matters of agreement and decision. [Cf. Lewis (1969).]

With respect to the nature of truth, therefore, it appears as though the analysis initiated by Austin carries us along on the right path. His notion of "convention", of course, needs to be exchanged for the notion of "habituation" in order to secure a theory of truth for individual idiolects as opposed to regional dialects and to community languages. Perhaps even more important to observe from the present perspective, however, is the reason why truth itself should be worth having. When our actions are based upon beliefs that are true, they are in those respects appropriately guided. When they are based upon beliefs that are false, they are in those respects unfortunately misguided. As an objective property relating language to the world, therefore, we need to have an objective conception of truth to explain in part why we succeed when we do succeed and why we fail when we do fail. As a subjective property relating our beliefs to our behavior, however, we benefit from having a subjective counterpart to that objective conception to explain the same phenomena from our own subjective point of view. And from this point of view, it makes a lot of sense to say that our beliefs are true when they are appropriate to guide our behavior.

5. VARIETIES OF KNOWLEDGE

With Chapter 4 as background, we are now in a position to take a closer look at a series of issues in AI with the resources required to clarify and illuminate what is going on from the perspective of the theory of knowledge. It will turn out that the problem of capturing common-sense knowledge has several different facets, not least of which are arriving at decisions as to the kind of knowledge that common sense is supposed to represent, on the one hand, and the extent to which capturing that knowledge by means of a computer program would be a desirable result, on the other. During the course of this discussion, we shall have occasion to draw distinctions between several different varieties of knowledge, including common-sense knowledge, defeasible knowledge, scientific knowledge, and conversational knowledge. Perhaps the most surprising result of this analysis turns out to be the discovery that a relatively obscure requirement that affects the truth of laws in science promises to yield a uniform framework for dealing with them all.

The conception of truth as appropriate belief suggests a crucial link with potential behavior that tends to motivate the search for knowledge. For it seems difficult to deny that we receive appropriate guidance from our beliefs when they are true and inappropriate guidance when they are false. And this in turn suggests that the more we know, the better prepared we are to contend with whatever comes our way. If we possessed knowledge about the future upon which we could absolutely rely (because of its infallibility), we might be able to anticipate exactly what problems we will face and what we will not. Knowledge of this kind, however, appears to be reserved for beings other than human. Our best strategy in preparing ourselves for the future, therefore, would seem to be the acquisition of knowledge that is broad in scope and rich in detail concerning those problems we are most likely to encounter as a means toward the attainment of our goals.

No one would be inclined to think that belief alone would be sufficient to generate these benefits. No one would suggest that the more we *believe* the better prepared we are to deal with whatever comes our way. No one would maintain that we receive appropriate guidance from our beliefs *just because we hold them.* And no one would contend that our best strategy in seeking to prepare ourselves for the future would be the acquisition of beliefs that are as *broad in scope and rich in detail* concerning the problems we believe we will encounter as a means toward the attainment of our goals. For the possession of a set of beliefs merely as a set of beliefs imposes no constraints at all with respect to consistency, closure, partial evidence, or complete evidence conditions. As sets of beliefs, one person's beliefs at one time are on a par with another person's beliefs at any other.

It is therefore fascinating to discover that Allen Newell has drawn attention to a conception of knowledge within AI which appears to have none of the virtues of "knowledge" and all of the vices of "belief" that we have elucidated above. That this is the case, however, might not be obvious upon preliminary consideration, for Newell employs familiar-sounding terms such as "knowledge" and "rationality" with unfamiliar senses, from the perspective of the theory of knowledge. Perhaps the most important element of his position assumes the form of a "behavioral law" intended to connect agents to situations and to goals, which he calls "the principle of rationality":

Principle of Rationality: If an agent has knowledge that one of its actions will lead to one of its goals, then the agent will select that action,

where "selecting an action" means performing that act [Newell (1981), p. 8]. Newell admits that this principle is not sufficient to determine the behavior of an agent in those situations when an agent confronts more than one action alternative, which might be addressed by adding "auxiliary principles".

There is a small problem with this principle as it has been formulated, I think, since it should make initial reference to "one of its *possible* actions", but that is easily repaired. There is a larger problem, however, that is not so easy to repair. If an agent needed money and *knew* that if he were to rob a bank (write a bad check, counterfeit a credit card) then he could get money, that does not mean that he *will* rob a bank (write a bad check or counterfeit a credit card). This is not a matter of subtlety, moreover, but one of morality. Yet it plainly suggests that Newell's principle is not true. The possibility remains, however, that one of his auxiliary principles concerning alternative action, for example, might be of help here [Newell (1981), p. 8]:

Equipotence of Acceptable Actions: For given knowledge, if action *A*1 and action *A*2 both lead to goal *G*, then both actions are selected.

While explicit reference to an agent and his goals is suppressed, it is taken for granted goal *G* is one of that agent's goals. In this case, for an action to be selected simply means that it becomes "a member of a candidate set of actions, the *selected set*, rather than being the action that actually occurs".

An issue that arises now is that the mere potential that some possible action might lead to a desired goal surely does not mean that it has to belong to such a set. The influence of morality continues to be excluded from consideration here, where *an action is unethical* for such an agent when he rules it out as an unacceptable means toward an end on moral grounds. Unethical actions are means toward ends which an agent is unwilling to adopt. This problem simply generalizes on one we have previously identified. A more significant result, therefore, is the realization that what an agent will or will not do depends not only on his motives, ethics, and beliefs but also upon his abilities, capabilities, and opportunities. The theory of rationality of action cannot succeed unless it is rooted in an adequate theory of action.

A quite different and equally damaging difficulty arises at this juncture in the form of Newell's conception of knowledge, which is based on his principle of rationality itself. For Newell suggests the following as its definition:

(I) knowledge (of an agent) =df whatever can be ascribed to the agent, such that its behavior can be computed according to the principle of rationality.

As he observes, this conception is intended to relate an agent's inner states to his external behavior and to characterize knowledge *functionally* in terms of what it does rather than *structurally* in terms of physical things and their properties and relations to other things [Newell (1981), pp. 9-10]. Whatever its other virtues, however, this definition not only suffers from the problems that attend the principle of rationality and the equipotence of acceptable actions but also invites dispute by ignoring the role of warrants in knowledge.

The relationship between Newell's conception of rationality and his definition of knowledge deserves critical scrutiny. For his principle of rationality is formulated in terms of knowledge and his definition of knowledge is formulated in terms of his principle of rationality. The circularity that is evident here might or might not be objectionable, depending upon how these notions are otherwise unpacked. But notice that there is nothing about Newell's use of the term "knowledge" that requires an agent's knowledge to be warranted or true! And there is nothing about his use of the term "rationality" that requires an agent's knowledge to be warranted or true! As a consequence, although Newell's use of these terms creates the impression that he is talking about *knowledge* and *rationality* as those terms occur within the theory of knowledge, nothing, in fact, could be further from the truth. For the strongest requirement that Newell's conceptions satisfy is the existence of beliefs.

In order to appreciate the distance separating Newell's notion of knowledge from epistemic analyses, observe that the refined conception of knowledge elaborated in Chapter 4 imposes four distinct requirements upon a set of beliefs of an agent z at a time t in order to qualify as "knowledge". These include two deductive requirements, (CR-1) and (CR-2), that insist upon closure and consistency, in conjunction with two evidence requirements, (CR-3) and (CR-4), imposing partial and complete evidence conditions. Taken altogether, no doubt, they represent a normative idealization that human beings exemplify only partially or to some degree. But suppose they were entirely abandoned. It would then be the case that inconsistent beliefs (which could not possibly be true) and fantastic speculations (for which there exist not a shred of evidence) could nevertheless qualify as "knowledge". Such crucial differences between "knowledge" and "belief" deserve to be acknowledged.

A more explicit formulation of Newell's "principle of rationality" in the sense of belief that was intended, therefore, would be something like this:

*Principle of Rationality**: If an agent believes that one of its actions will lead to one of its goals, then the agent will select that action,

where it should be obvious that this principle is very difficult to defend, especially because motives and beliefs are not the sole determinants of action. Thus, in the case of his definition of "knowledge", the following seems to fit:

(II) knowledge (of an agent) =df whatever can be ascribed to an agent, such that its behavior can be computed in accordance with its beliefs.

It ought to be evident that this is also a dubious principle of rationality for human actions. Neither of these notions implies that "knowledge" needs to be justified or that "rationality" may involve more than motives and beliefs.

Newell provides a partial response to issues such as these by remarking that the term "knowledge" occurs in the theory of knowledge with the very different meaning implied by *justified true belief*; but he dismisses this notion as irrelevant to artificial intelligence on the mistaken assumption that, in order for anything to be known, it must be *known with certainty* [Newell (1981), p. 18]. He acknowledges the similarities between his conceptions of knowledge and rationality and "the intentional stance" that Dennett has proposed, according to which motives and beliefs may be ascribed to anything at all when its behavior can be predicted and explained "as if" it had those motives and beliefs. As we previously discovered, treating a thing *as if* it had these specific properties does not mean it actually has them, where it seems evident that Newell's own conceptions are subject to all of the same objections that were raised against Dennett's conception back in Chapter 1.

The point of this exercise is not merely to explain that notions like these have sometimes been used with very different senses in epistemology and in artificial intelligence, which is very clearly true. What Newell means by "knowledge" and "rationality" are simply not their standard meanings, for which philosophy may claim precedence by more than 2000 years. What is even more important to discern is that the framework that Newell represents is a very different conception than is reflected by the theory of knowledge as that discipline has traditionally been pursued. Indeed, from the perspective provided by Newell's account, the relationship between the theory of knowledge and the theory of knowledge representation takes a new form as the theory of belief and the corresponding theory of belief representation:

	Theory of Belief:	*Theory of Belief Representation*:
Objective:	To discover the kinds of things that can be believed	To discover the ways in which beliefs can be represented
Assumption:	That there are various things that can be believed	That there are kinds of beliefs to represent

Fig. 14. Dependency Relations II.

As long as the notion of "belief" is being employed in more or less the same senses by theories of knowledge in epistemology and by theories of knowledge in Newell's AI sense, comparison with Figure 9 of Chapter 4 ought to indicate that these are different conceptions of knowledge representation.

The discovery that the term "knowledge" occurs in different disciplines having different senses, of course, does not necessarily mean that either of those senses ought to be revised or be abandoned. [Rapaport (1986) and Levesque (1988) have made similar observations.] Nevertheless, I want to suggest that Newell's sense of "knowledge" is a very weak notion that does not adequately clarify and illuminate the problems of knowledge that arise with respect to the problems of knowledge representation. The appropriate measure of their respective merits does not merely consist of the extent to which they can contribute toward the solution of problems within their original domains. For it is my contention that the evidence supports the traditional theory of knowledge over the nontraditional theory of knowledge in relation to their respective capacities to deal with problems arising in AI. And if I am right in thinking this is true, then it must be the case that the resources provided by the traditional theory of knowledge (as represented by Chapter 4) can contribute toward the solution of major difficulties in AI.

ORDINARY KNOWLEDGE

A good place to begin, I think, is by drawing a distinction between ordinary knowledge and scientific knowledge, where ordinary knowledge satisfies the standard conception of knowledge as warranted, true belief, while scientific knowledge satisfies only the conditions for warranted belief. One basis for drawing this distinction is that the subjects of ordinary and of scientific knowledge appear to be quite different, where ordinary knowledge tends to be limited to singular sentences about ordinary things and to generalizations expressed in ordinary language. Scientific knowledge thus extends to singular sentences about extraordinary things and to generalizations that require technical language. We shall presume that ordinary knowledge is the property of an *individual* knowing thing z, while scientific knowledge is the property of a *community* of knowing things Z, which might have as few as only a single member. Later we shall return to the significance of this difference. While this distinction suffers from the vagueness that comes with appeals to "ordinary" and "extraordinary" things, we may assume that the boundary intended is sufficiently clear for our present purposes (and pursue these issues further when that turns out to be required). [Cf. Fetzer (1981), Part I.] The differences that are intended here, therefore, might be represented by the following sets of conditions for "ordinary" and for "scientific" knowledge:

(III) (a) an individual *z* knows *p* (in the sense of *ordinary knowledge*) if and only if (i) *z* believes that *p*, (ii) *p* is the case, and (iii) *z* is justified in believing that *p*; whereas,

(b) a community *Z* knows *p* (in the sense of *scientific knowledge*) if and only if (i) *Z* believes that *p*, and (ii) *Z* is justified in believing that *p*, whether or not *p* is the case.

Thus, the omission of the truth condition from definition (b) means that any changes that occur in the state of scientific knowledge as a result of the accumulation of evidence in the form of observations and experiments (which might qualify as ordinary knowledge, depending on their character) do not thereby render earlier states of scientific knowledge as never having been "scientific knowledge", but rather as the scientific knowledge of their time. The reasons for this view are not beyond debate [Fetzer (1981), pp. 14-16].

Common Sense. The nature of rationality – i.e., rationality of action – has also become important to AI from the perspective of problems involved in representing *common-sense knowledge.* John McCarthy, for example, tends to formulate the problem as that of developing a formal theory of "the kind of means-ends analysis used in daily life" [McCarthy (1968), p. 410]. He is concerned with capturing how we come to draw conclusions and arrive at decisions on the basis of what we know, with special concern for the often fragmentary and incomplete standing of our knowledge. McCarthy insists that the knowledge reflected by common sense is different in kind from the knowledge reflected by the exact sciences, such as theoretical physics. Thus,

> Our system is not intended to supply a complete description of situations nor the description of complete laws of motion. Instead we deal with partial descriptions of situations and partial laws of motion. Moreover the emphasis is on the simple qualitative laws of everyday life rather than on the quantitative laws of physics. As an example, take the fact that if it is raining and I go outside I will get wet. [McCarthy (1968), p. 411]

In the pursuit of this objective, therefore, he advances a "logic of situations".

There are several problems with McCarthy's conception, which are shared by various other conceptions of this problem. The first appears to be the assumption that there *is* such a thing as "the kind of means-end analysis used in daily life". Those who till the field of the theory of decision, for example, have separated maximizing policies from satisficing policies from cost-benefit policies, and these are merely three among a rather populous set of alternative decision-making policies. [For an introduction to principles available within this domain, see, for example, Michalos (1969), Part II.] Given the importance of ethical commitments in affecting human behavior, on the one hand, and the existence of alternative decision-making policies, on the other, there are no reasons to believe and many reasons to doubt that one and only one distinct decision-making policy is ever employed in daily life.

The second difficulty is McCarthy's willingness to embrace partial descriptions of situations and incomplete specifications of the laws that govern this domain. Taking advantage of a distinction that we have previously discussed, this means that he wants to deal with human behavior by treating human beings as *open* systems rather than as *closed* systems, where open systems are subject to the influence of causal factors other than those that have been explicitly specified. The complexity of human behavior is such, however, that the omission of even a single relevant feature of the beliefs, motives, ethics, abilities, capabilities, and opportunities that influence the behavior of a human being could make an enormous difference in explaining and predicting his behavior. By adopting the policy of treating human behavior as the product of open systems in this sense, McCarthy introduces not just the likelihood but a *de facto* guarantee that the generalizations constituting the core of common-sense knowledge will be subject to exceptions.

As an example, consider McCarthy's own illustration cited in the passage quoted above – "if it is raining and I go out I will get wet" – which he refers to as "a fact". Notice, in particular, that if it is raining and I go out *but wear my raincoat* I need not get wet. If it is raining and I go out *but stay under the veranda* I might not get wet. If it is raining and I go out *but remain beneath the tent set up on the lawn for tomorrow's wedding* I will not get wet. And this is a matter of information or knowledge that appears to qualify as "common sense". So, far from being a "fact", McCarthy's example is not even true, an inevitable result of treating only a few of the factors that can make a difference to the occurrence of an outcome as though they were complete. The point, of course, is not that an author made mistakes in a paper twenty years ago, but that false generalizations of this kind are not beneficial to AI.

The problem appears to be that McCarthy and Newell are treating crude principles and "rules of thumb" that are obviously amenable to exception as as though they were exceptionless generalizations. Even if they were true, they would reflect coincidental relations that do not have to happen. It is sometimes the case that if it is raining and I go outside I will get wet (say, whenever I go without my raincoat, leave the veranda, and do not stay under a tent set up over the lawn for tomorrow's wedding – unless something else occurs that has the effect of keeping me from getting wet, which might be a precaution that I take or something else beyond my control, etc.). And it sometimes happens that only one idea comes to mind in the endeavor to realize a person's goals, where he does not rule it out on moral grounds and has the abilities, capabilities, and opportunities to pursue it, in which case it may happen that this is what he does. But not always or necessarily, etc.

Other illustrations are not difficult to discern, especially in domains that emphasize the importance of "exceptions to the rule", including the area of ethics. We would normally assume that telling the truth and obeying our parents are the sorts of things that we ought to do. But suppose that little Billy has hidden in our basement because a local bully is searching for him everywhere. If the bully arrives on our doorstep with baseball bat in hand and asks for Billy's whereabouts, it would be a mistake to suppose that he ought to be told because "lying is wrong". When a youngster's parents tell her to shoplift some drugs from the local pharmacy, it would be a mistake to suppose that she should do what she is told merely because "children ought to obey their parents". When "rules of thumb" are interpreted as generalizations without exceptions, their role as "guidelines" has been misunderstood.

The pattern of reasoning involving crude generalizations and "rules of thumb" that McCarthy wants to implement by means of a program reflects a form of common sense that can be subject to defeat by the acquisition of new evidence. There are similarities, but his position differs in motivation from one that has been offered by Marvin Minsky in formulating (what has come to be called) his "Dead Duck Challenge" [Nilsson (1983), p. 10]. As Nils Nilsson has described it, Minsky questions whether formal schemes for the representation of knowledge (including especially first-order logic) are appropriate frameworks for representing knowledge, because they can lead to faulty inferences when generalizations (such as "All birds can fly") are conjoined with real-world exceptions (such as dead ducks). Thus, for example,

(IV) All birds can fly
Zap is a bird

Zap can fly

is an example of a presumably valid inference. Yet if Zap is a dead bird, the conclusion of this inference, although it follows validly, turns out to be false.

The problem with Minsky's "challenge" arises from insufficient attention to the difference between *validity* and *soundness* as properties of deductive arguments. Example (IV) appears to be valid, since it is difficult to see how, if its premises were true, its conclusion could be false. When the conclusion of a valid argument is false, however, it cannot possibly be the case that all of its premises are true. This difference separates *valid* or *sound* deductive arguments from *proper* or *correct* inductive arguments: the conclusions of proper or correct arguments can still be false when their premises are true. In the case of Minsky's "Dead Duck Challenge", therefore, if we assume the argument is intended to be deductive (diagrammatically using a single line between premises and conclusion to mark a deductive relation), then if it *is* a valid argument, its premises cannot all be true when its conclusion is false. So if it is true that all birds can fly and it is true that Zap is a bird, then the possibility that Zap cannot fly is ruled out by the validity of this argument.

Thus, if the conclusion is false but the argument is valid, then it must be the case that Zap cannot fly. So either Zap is not a bird or it is not the case that all birds can fly. If "bird" as it occurs in the first premise has the same meaning as "bird" as it occurs in the second premise, then since Zap is *dead* it must be possible to be something that is both dead and a bird. But if this is possible, then the first premise must be *false*, since then it is not true of all birds (dead or alive) that they can fly. Analogously if "bird" as it occurs in the second premise has the same meaning as "bird" as it occurs in the first premise, then since all birds can *fly* but Zap is dead, it cannot be possible for Zap to be a bird. But if this is impossible, then the second premise must be *false*, since it is untrue of Zap that it is still a bird (when it's dead).

An alternative construction of this example, however, would envision the argument as inductive rather than as deductive and reconstruct it as follows (marking it inductive by a double-line between its premises and conclusion):

(V) Most birds can fly
Zap is a bird
═══════════
Zap can fly

assuming, as before, that Zap is among the dear departed. In this case, there is no question of validity, since the argument only claims to be proper. And the propriety of the agument is compatible with the falsity of its conclusion, even when its premises are true. But now, of course, its first premises says something about *most* birds rather than about *all* birds, in which case there is no reason to misunderstand that, even if the vast majority of birds can fly, this provides no guarantee that Zap specifically has to be among them.

A simple measure of partial support (which may not endear itself other than as an accessible illustration) might be employed to exemplify the sort of inferential situation encountered by inductive reasoning in cases of this kind. Suppose, for example, that the degree of partial support for a conclusion in arguments of this form must have the same numerical value as the relative frequency for the occurrence of its attribute (such as *can fly*) within a corresponding reference population (such as that of *birds*), as follows:

(VI) m/n birds can fly
Zap is a bird
═══════════ $[m/n]$
Zap can fly

where "m/n" in the first premise reflects the relative frequency with which things that are birds are things that can fly, and "m/n" between brackets reflects the degree of partial support that those premises confer upon that conclusion. Clearly then, unless $m = n$, the conclusion could be false when both of its premises are true, yet the argument could still be inductively proper.

The example could be redone to make explicit mention of Minsky's dead ducks rather than dead birds, but the upshot would be the same. In either event, the Dead Duck Challenge trades upon mistakes, which can be one or another of two distinct kinds. A mistake of the first kind results from ambiguity if words that occur in some premises occur with different meanings in other premises. This appears to be the case with respect to example (IV), of course, where the argument is interpreted as deductive. In this case, the word "bird" turns out to be ambiguous, which violates the requirement of a uniform interpretation. A mistake of the second kind occurs when an argument that is intended to be inductive (or deductive) is interpreted as deductive (or inductive) instead. This would be the case if the original argument used the sentence, "All birds can fly", as a casual expression of the meaning, "Most birds can fly", as displayed by example (V). In this case, the premise "All birds can fly" turns out to be false, but it is subject to reinterpretation.

Minsky's mistake appears to be an error of the first kind, since he almost certainly intended to assert that *all* birds can fly rather than that *most* birds can fly. But these problems are interrelated, because the truth value of the claim that all birds can fly depends upon the meaning of the word "bird" itself. Without semantical clarification in the form of a meaning stipulation as to whether dead things can properly qualify as birds, the problem he poses trades upon an equivocation. Once this semantical stipulation has been rendered, however, it becomes apparent that his position commits a blunder of one or another of the kinds I have explained. Minsky's Dead Duck Challenge thus hinges on an oversight by its proponents, for scrupulous adherence to the requirement of a uniform interpretation and explicit recognition of the characteristics of inductive and deductive arguments render it null and void.

Defeasible Reasoning. Given these results, it is intriguing to find there is at least one different interpretation that might be placed upon Minsky's position, which is that the Dead Duck Challenge represents a distinct and heretofore largely unexplored species of argument that is neither deductive nor inductive in kind. This position has been defined by Donald Nute as follows:

> We reason defeasibly when we reach conclusions that we might be forced to retract when faced with additional information. I contrast this with both invalid deductive reasoning and inductive reasoning. This reasoning is defeasible, but its defeasibility is not because of incorrectness. Nor is it ampliative as is inductive reasoning. It is the "other things being equal" reasoning that proceeds from the assumption that

we are dealing with the usual or the normal case. Conclusions based on this kind of reasoning may be defeated if we find that the situation is not usual or normal. [Nute (1988), p. 251]

Indeed, from this point of view, it seems plausible to suppose that Minky's Dead Duck Challenge, far from reflecting a blunder, actually represents the implicit discovery of a new species of "defeasible reasoning" in Nute's sense.

The difficulty with this position, as Terry Rankin has observed, is one of establishing that reasoning of this kind is not better understood as a special case of inductive reasoning instead [Rankin (1988)]. For if this is a kind of "other things being equal" (or *ceteris paribus*) form of reasoning, it should be possible to formalize any argument of this kind more or less as follows:

(VII) *Ceteris paribus*, *X*s are *Y*s
This is an *X*
- - - - - - - - - - - - - - -
This is a *Y*

where the series of dashes separating premises from conclusion reflects uncertainty over the kind of reasoning at stake here. In application to the example we have examined above, a defeasible argument would look like this:

(VIII) *Ceteris paribus*, birds can fly
Zap is a bird
- - - - - - - - - - - - - - - - -
Zap can fly

But of course the *ceteris paribus* clause itself covers any number of special cases: birds that are dead, birds that have broken wings, birds that are too young, etc. In asserting that Zap is a bird and drawing the conclusion that Zap can fly, it certainly appears as though an inductive risk is being taken.

Consider, especially, that the conclusion that Zap can fly might be false even when these premises are true. The requirement of a uniform interpretation, no doubt, could be invoked to rule out the possibility that Zap is dead, so the word "bird" does not occur with one meaning in the first premise and another in the second. But that does not eliminate the other possibilities that Zap might have a broken wing or be too young to fly, etc. This means that an argument of this kind would be valid only if it were the case that the *ceteris paribus* clause itself could be assumed to be satisfied, thus:

(IX) *Ceteris paribus*, birds can fly
Zap is a bird
Ceteris paribus (i.e., all other conditions are equal)
- -
Zap can fly

in which case the conclusion could not be false if its premises were true [cf. Rankin (1988), pp. 305-306]. The argument would no longer be defeasible.

Hence, however tempting it may appear to view defeasible arguments as a novel species of argument that is neither inductive nor deductive, I think the weight of the evidence does not support such an interpretation. Indeed, from this point of view, Nute's interpretation of these arguments appears to confront a dilemma. For if satisfaction of the *ceteris paribus* clause is made explicit, such an argument appears to be demonstrative, non-ampliative, and additive, which implies that it assumes a deductive form like the following:

(X) *Ceteris paribus*, birds can fly
Zap is a bird
Ceteris paribus (i.e., all other conditions are equal)
____
Zap can fly

But if satisfaction of the *ceteris paribus* clause is not made explicit, such an argument appears to be non-demonstrative, ampliative, and non-additive, which implies that it assumes an inductive form like the following instead:

(XI) *Ceteris paribus*, birds can fly
Zap is a bird
========================
Zap can fly

Surely satisfaction of the *ceteris paribus* clause either is made explicit or it is not, which implies that these arguments are either deductive or inductive. Thus, as Rankin maintains, these seem to be the only available alternatives.

If these considerations are well-founded, however, then there does not appear to be any special classification of arguments that are defeasible but neither deductive nor inductive. Leaving fallacies aside, there are just two broad varieties of arguments, one species of which is intended to be truth-preserving, another species of which is meant to be knowledge-expanding, where the basic notions that define these species continue to be as follows:

	Right Form:	*Right Form and True Premises:*
Deductive Argument:	Valid	Sound
Inductive Argument:	Proper	Correct

Fig. 15. Species of Arguments.

It thus seems appropriate to conclude that defeasible reasoning is indeed a special case of inductive reasoning, in general, which serves to explain why arguments involving crude generalizations and rules of thumb appear to be non-demonstrative, ampliative, and non-additive, as we have found above.

SCIENTIFIC KNOWLEDGE

The types of knowledge discussed by McCarthy and by Minsky (and also by Newell and by Nute) do not exhaust what have been taken to be cases of common-sense knowledge. Perhaps the most remarkable instances concern (what tends to be called) *common-sense* (or "*naive*") *physics* [Hayes (1978), (1985)]. Patrick Hayes has championed the idea that AI ought to devote itself to formalizing our naive (or common-sense) worldview, no matter how complex an undertaking that might become. The success of a theory of this kind is supposed to be measured by the extent to which it provides a family of tokens and relations that yields the inferences that we draw in daily life:

> People know, for example, that if a stone is released, it falls with increasing speed until it hits something, and there is then an impact, which can cause damage if the velocity is high. The theory should provide tokens allowing one to express the concept of releasing a stone in space. And it should then be possible to infer from the theory that it will fall, etc.: so there must be tokens enabling one to express ideas of velocity, direction, impact, and so on. [Hayes (1985), p. 5]

And this, indeed, is an intuitively appealing conception, whose fulfilment he seeks to advocate as one of the most important objectives for AI to pursue.

Nonetheless, several aspects of this position are strikingly problematical. The first, no doubt, would fascinate historians of science, for what probably qualifies as the greatest system of naive physics the world has ever known was formulated during the 4th C. B.C. by Aristotle. Aristotle employed the notion of "natural place" to account for the motion of physical things, with a concentric sphere model of the planets and the stars that persisted and endured through much of the Middle Ages. Indeed, even more impressive to AI, from Hayes' perspective, ought to be his theory of causation in terms of formal, final, material, and efficient kinds of causes, and his system of categories that are intended to capture the kinds of properties that can be predicated of things, which include: substance, quantity, quality, relation, place, time, situation, state, action, and passion [cf. Ross (1956) and Jaeger (1960)].

Aristotle's theory of naive physics, which is rooted in what can be observed with the naked eye, is one of the fascinating achievements of the human mind. His conception of the nature of causation appeals greatly to intuition, and his categories are enormously plausible from the point of view of common sense. Indeed, Aristotle's theory of predication – especially, his theory of essential predication – introduces distinctions between "essential" and "accidental" properties that compare favorably with most current accounts of semantic networks, which reflect relatively feeble attempts to come to grips with issues that Aristotle understood quite clearly. Yet it seems stunning to think that one of the foremost objectives of artificial intelligence should be taken to be the attempt to capture any theory of naive physics of this kind. The development of modern physics, after all, has depended upon overcoming Aristotelian accounts, including his teleological conception of explanation. [For accessible introductions to this area, see Kuhn (1957) and Cohen (1985).]

A second line of reasoning that raises certain doubts concerning whether or not common-sense physics should be viewed as an AI project worthy of pursuit arises from the generalizations that would constitute its core. Consider Hayes' own illustration, for example: that if a stone is released, it falls with increasing speed until it hits something, when there is an impact that can cause damage if the velocity is high. Hayes appeals to intuition, and this generalization on its face seems compatible with common sense. Generalizations such as this one, however, have innumerable exceptions.

What if a stone is released, but it is tied with string to a large helium balloon: would it fall with increasing speed? Suppose the stone is very, very small and as it falls it increases in speed: if it impacts a window pane, will it cause damage? or if it impacts a concrete bunker, will it cause damage? What if a stone is released by an astronaut in orbit around the Earth: would it fall with increasing speed? with decreasing speed? with no speed at all? What's going on?

The problem, I surmise, is that *common-sense knowledge* is not well-defined as a fixed body of knowledge or beliefs. What is intuitively adequate to one person at one time may not be intuitively adequate to someone else at that same time, or even to that same person at some other time. It is all a matter of how much we know and how long we have known it. When we know a great deal about any subject matter, our views about that material become "second nature", "instinctive beliefs", "obvious and trivial, but true". But there simply does not exist any specific set of sentences that ought to be included within a program of this kind. Babes in arms, small children, teenagers and young adults, scholars and nitwits, the aged and the infirm, all enjoy beliefs that are sufficiently familiar to them as to qualify as nothing but "common sense". But that hardly shows that there is something here worth knowing, much less formalizing and implementing in the form of a program.

And this problem is ubiquitous. It occurs in many domains besides that of physics. Consider, for example, ordinary human behavior. We now know that an adequate theory of rational action presupposes an adequate theory of action, whose outlines have become more or less clear. Still, we may ask if there is a common-sense theory that could serve as the foundation for a computer program that would successfully explain and predict our behavior. We shall explore this problem again when we turn to scripts and frames as modes of knowledge representation. For the moment, however, it might be worthwhile to reflect upon *widely-held beliefs* such as those uncovered by Robert Lynd's classic study of American society. Consider these examples:

(XII) (1) "Everyone should try to be successful. *But*: the kind of person you are is more important than how successful you are."

(2) "Honesty is the best policy. *But*: Business is business, and a businessman would be a fool if he didn't cover his hand."

(3) "The American judicial system insures justice to every man, rich or poor. *But*: A man would be a fool not to hire the best lawyer he can afford."

(4) "Religion and 'the finer things of life' are our ultimate values and the things all of us are really working for. *But*: A man owes it to himself and to his family to make as much money as he can." [Lynd (1948), pp. 60-62]

If this is "common-sense knowledge", we may well ask, "What good is it?"

The problem, I suspect, is double-edged, since common-sense knowledge appears to be an inconsistent body of beliefs that permits the prediction or the explanation of virtually any behavior by anyone at all. Consider, for example, the case of a young husband whose wife visits her parents for a long weekend by herself. If he were to take advantage of the situation by seeing a former girlfriend on the sly, we could "explain" his behavior by appealing to the common-sense belief, "Out of sight, out of mind". But if instead he remained at home, cleaning the house and doing the dishes to surprise her upon her return, we could "explain" that, too, by appealing to the common-sense belief, "Absence makes the heart grow fonder". Thus, the problems with inconsistent beliefs are the same as the problems with inconsistent sentences: they not only cannot all be true, but they can be used to "explain" anything!

If the objective is to discover what people actually believe, of course, the existence of inconsistent beliefs should come as no surprise. But this fact of life should create at least some doubts over the importance and desirability of discovering what they are and packaging them in the form of a program. If the objective is to discover what people ought to believe, by contrast, the existence of inconsistent beliefs can serve a heuristic role. For in this case it would be useful to reflect upon what "common sense" reports in an attempt to discover whether or not some consistent and illuminating sets of beliefs underlie these linguistic manifestations. It depends on whether we regard "common sense" as a source of *conclusions* or as a source of *evidence*. For the project of sorting things out for the purpose of discovering a consistent and illuminating theory (no matter whether in physics or in psychology), if such is at hand, requires not common sense but expertise within a domain.

What common sense can provide by now ought to be clear. For all of the examples that we have considered – from Newell to McCarthy to Minsky to Nute and to Hayes – involve crude generalizations that can function, at best, as "rules of thumb" that have numerous exceptions. And this circumstance, I suspect, generates an implicit paradox. For if common-sense knowledge is taken at face value, the generalizations it provides turn out to be untrue (as "rules of thumb" with numerous exceptions); but when the "generalizations" it provides are qualified by the numerous exceptions required to make them true, they no longer count as common sense but rather as expert knowledge. And the features that distinguish experts from amateurs within a field or a domain are precisely those that separate programs worth writing from programs without value. When all is said and done, however, surely that is exactly what we should have suspected all along. For the beliefs possessed by amateurs (about flying birds, falling stones, or human behavior) are almost never equal in value to the knowledge possessed by experts (in ornithology, in physics, or in psychology). The bedrock of knowledge upon which AI is destined to flourish, it seems, is not common sense but scientific knowledge.

Laws of Nature. The aim of empirical science, generally speaking, is to discover laws of nature that might be systematically employed for the purpose of predicting and explaining the events that occur during the course of the world's history. There is a division of labor, whereby the various special sciences pursue the laws of their respective domains: physics, the laws of physics; chemistry, the laws of chemistry; biology, the laws of biology; and so forth. Empirical sciences are both theoretical and experimental in their methods and techniques. The theoretical part of empirical science involves making guesses (using theories and models) as to what the laws of nature are, while experiments and observations help to sort out the good guesses from the bad (providing ways to test them). In general, therefore,

(XIII) Sciences are both:
1) theoretical, using
   a) theories, and
   b) models; and
2) empirical, using
   a) observations, and
   b) experiments;

where "observations" involve perceputal inference, while drawing conclusions from experiments often combines inductive with deductive reasoning.

Sometimes theories are distinguished from laws of nature, which may be appropriate when theories are cast as formal systems to which empirical interpretations are assigned. There are, however, at least three distinct alternative conceptions of the nature of theories in science. The first takes them to consist of an *abstract calculus* combined with an *empirical interpretation*, and it has come to be known as "the standard conception" [Braithwaite (1953) and Hempel (1965)]. The second takes them to be *theoretical definitions* that may be used to make *empirical claims*, and it is often referred to as "the semantic conception" [Suppes (1967) and Sneed (1971)]. The third takes them to be sets of *lawlike sentences* conjoined together with *meaning postulates* (as definitional sentences that might be either "partial" or "complete") which apply to a common domain, to which we shall refer as "the Campbellian conception" [cf. Campbell (1920) and Fetzer (1981), pp. 156-161]. Only the third, however, represents an appropriate relationship between theories on the one hand and laws of nature on the other [Fetzer (1985b), pp. 232-236].

Sometimes models are distinguished from theories, but sometimes not. The phrase "causal model", which will become important especially in relation to the idea of program verification as it is examined in Chapter 8, has been bestowed upon entities as diverse as *scientific theories* (such as classical mechanics and relativity theory), *physical apparatus* (such as arrangements of ropes and pulleys), and *operational definitions* (such as that I.Q.s are what I.Q. tests test). Indeed, different disciplines tend to generate their own special senses [cf., for example, Hesse (1966), Heise (1975), and Suppe (1977)]. Interpreted as physical things, "models" become good models when (i) there is a one-to-one correspondence between the parts of the model and the parts of the thing being modeled, and (ii) the parts of the model happen to be arranged so as to preserve the relations that obtain between the parts of the thing being modeled. Precisely which aspects of the thing being modeled have to be reflected by the model itself, however, is not fixed once and for all, but rather depends upon and varies with the point of view intended.

On all of these accounts, theories are linguistic entities. For the standard conception and the Campbellian conception, they are true when the sentences of which they are composed (representing properties of the world) are true. By contrast, for the semantic conception, theories themselves are true by definition, while the empirical claims that may be made using them are, in turn, true when the world has the properties they attribute to it (as synthetic, rather than analytic, sentences). But the Campbellian account is the one that supports the most appropriate account of the structure of science:

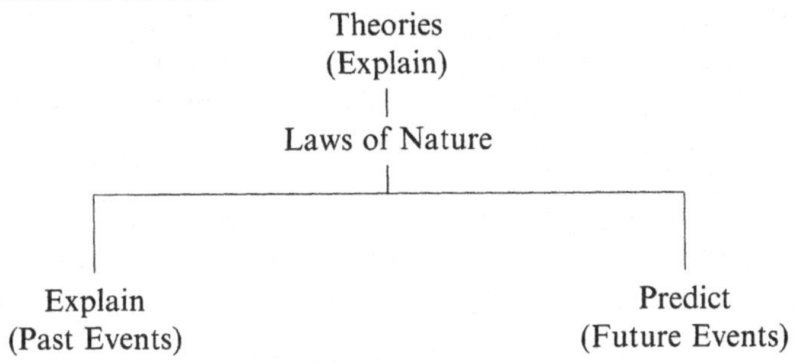

Fig. 16. The Structure of Science.

Thus, theories are used to explain laws, but laws in turn are used to explain and to predict the occurrence of singular events during the world's history.

Let us therefore adopt that account here. Then, when a theory happens to be true, it can be relied upon for the purposes of explanation and of prediction. But there are no built-in guarantees that theories that have been accepted as true might not turn out to have been mistakes. The principal reason this should be the case naturally results from the intersection of the nature of laws as the objects of inquiry and the nature of scientific inference as the method of inquiry. For laws of nature make claims about every actual or potential instance of a certain reference property, asserting that any thing that ever has been, is now, or will be an instance of that property (in the past, the present, or the future) must also be an instance of an attribute, precisely because that attribute is a "permanent property" of that reference property. And this provides an ontological justification for subjunctive and for counterfactual conditionals (regarding what would happen or would have happened if something were to be or had been an instance of that property).

As we discovered in Chapter 4, there is a fundamental difference between permanent and transient properties, which is relative to some presupposed reference property. In the case of permanent properties, there is no process or procedure, natural or contrived, by means of which something that has that reference propery could lose that attribute without also losing that reference property, even though its possession of that attribute is not true as a matter of definition alone [Fetzer (1981), Part I]. In the case of transient properties, however, there are procedures, either natural or contrived, by means of which something could lose that attribute without also losing that reference property. Things that are lead (if "lead" is defined in terms of its atomic number 82), for example, have their melting points and boiling points and malleability as permanent properties, but their size, their shape, and their current market value (when purchased in quantity) are transient.

Laws of nature, therefore, are clearly very different entities from laws of society. Consider, for example, a few very simple instances of each variety:

(XIV) (1) laws of nature include:
 (a) unlike magnetic poles attract,
 (b) copper is a good conductor, and
 (c) matter expands when heated;
(2) laws of society include:
 (a) 65 mph speed limits,
 (b) no trespassing areas, and
 (c) legal ages for drinking.

Laws of society are products of government. They are passed by a legislature, interpreted by the judiciary, and enforced by the executive branch of a government. Laws of nature, by contrast, require no law-giver and need no enforcement. Although laws of society can be changed and can be violated, laws of nature cannot be violated and cannot be changed. Our beliefs about which sentences representing laws of nature are true might change as new theories are advanced and as the evidence available – especially by means of technology, such as telescopes, microscopes, electron microscopes, radio telescopes, etc. – may change, but laws of nature themselves do not change. If a sentence concerning a natural law is true, it cannot be violated; and if a sentence concerning a natural law can be violated, then it cannot be true.

Laws of nature, unlike societal laws, apply in all places and at all times. Moreover, they may be used to explain their instances. Societal laws, however, cannot be used to explain their instances. Thus, even if everyone happens to obey a certain law, that could be explained only by explaining why each of those individuals happened to obey that law from time to time and from place to place, i.e., by reference to the motives, beliefs, ethics, abilities, capabilities, and opportunities that affected their behavior. Notice, in particular, that persons frequently violate laws of society, such as speed limits on highways, for what may or may not be very good reasons, such as getting a woman in labor to the hospital quickly or evading the police to avoid apprehension. Yet even if someone wanted to violate a natural law for what they took to be the very best reasons, say, by eating more but exercising less in order to lose weight, these efforts would be in vain, since it cannot be done. The laws of nature thus define the boundaries of what is physically possible.

Lawlike Sentences. The most important theoretical problem confronting philosophers of science, therefore, turns out to be the discovery of an adequate analysis of the nature of laws themselves. As properties of the world, no doubt, natural laws are physical, rather than abstract, entities. But the idea of "laws of nature" should not be construed so narrowly as to preclude the possibility of laws of behavior for humans and for machines, including, for example, pocket calculators and personal computers. The laws of the world are "physical" rather than "abstract" in the sense that they are part of the world's causal structure. Indeed, to be both more general and more precise, the world's laws define the world's nomic and causal structure, where laws of all kinds are subsumed by the term "nomic", which thus includes "causal laws" within its scope (while not all laws turn out to be causal in character).

From the point of view of language, however, laws of nature can also be viewed as linguistic entities, especially as *lawlike sentences* that happen to be true. This reduces the problem of understanding natural laws to understanding what it is to be a lawlike sentence (assuming, of course, that we already understand what it is for a sentence to be true). Some of the most influential and early work on this problem was advanced by Nelson Goodman, who suggested that lawlike sentences were those that could be confirmed by their instances [Goodman (1965)]. Unfortunately, accidental generalizations concerning transient properties are no less amenable to being confirmed by their instances than are lawlike generalizations concerning permanent properties. Thus, Goodman in effect provided an epistemic and pragmatic "answer" to an ontic and semantic "question", which has continued to confound the community of inquirers for almost twenty years [Fetzer (1981), Ch. 7].

The underlying difference between lawlike sentences and (merely) accidental generalizations emanates from the fact that laws preclude something a (mere) generalization permits, namely: the existence of processes or procedures, natural or contrived, by means of which an attribute, such as the color of a car, could be taken away from an instance of a reference property, such as a Volkswagen, without taking that reference property away as well, e.g., by repainting the car. As a linguistic manifestation of this ontological difference, therefore, subjunctive conditionals should be employed in lieu of material conditionals for these stronger claims, where a subjunctive '____ $\Rightarrow$ . . .' entails the corresponding material conditional '____ $\rightarrow$ . . .':

(XV) '$p \Rightarrow q$' entails '$p \rightarrow q$'.

That is, if a subjunctive is true, then the corresponding material conditional must be true, but not conversely. Indeed, the basic form of lawlike sentences is that of unrestricted, logically contingent, subjunctive generalizations. The difference between subjunctive generalizations and material generalizations, moreover, parallels a distinction drawn by Aristotle between the universal and the commensurately universal properties of things, but this difference is unpacked in terms of permanent-property relations instead of essential-property relations. *Universal properties* are properties that every thing of a certain kind happens to possess, while commensurately universal properties are properties that everything of a certain kind possesses "by its nature". The properties that something has "by its nature" are its *essential properties*, which are included in a proper definition of such a thing. While essential properties may be properties of things as those individual things, permanent properties are properties of things as instances of specific kinds. Permanent properties are *not* included in a proper definition of something. [Cf. Aristotle, *Posterior Analytics* 71a-74b, and *Metaphysics* 1028a-1032a.]

It is important to realize, however, that there is more than one way in which a subjunctive conditional (concerning what would occur if something were the case) might be true. For another class of subjunctive conditionals must be true simply as matters of definition. If the term "bachelor" means the same thing as the phrase "unmarried adult male" within some language framework, then it has to be the case that "If John were a bachelor, then he would be unmarried" is true. To differentiate the class of logically necessary truths from the class of lawlike sentences, therefore, it is appropriate to introduce a necessity operator '$\Box$ . . .' to identify logical truths, such that:

(XVI) (1) '$\Box(p \rightarrow q)$' entails '$p \Rightarrow q$', and
(2) '$p \Rightarrow q$' entails '$p \rightarrow q$'; however,
(3) '$p \Rightarrow q$' does not entail '$\Box(p \rightarrow q)$'.

It thus becomes easy to preserve a distinction between lawlike sentences and logical necessities. Lawlike sentences are (unrestricted) subjunctive generalizations that are not logically necessary. This reflects a difference in the variety of knowledge they represent when they are true. Lawlike sentences are synthetic in character, while logical necessities are analytic.

The subjunctive conditional, however, is not strong enough to represent causal connections, which not only assert that, say, if R were the case, then A would be the case (if x were gold, then x would have the melting point of 1063°C, for example) but also assert that *having R – invariably or probabilistically – causes (or "brings about") having A*. This additional meaning implies that causal conditionals must be distinguished from simple subjunctive conditionals which make lawlike, but not therefore causal, claims. Moreover, deterministic causal conditionals must be distinguished from indeterministic (or "probabilistic") causal conditionals, whose attribute outcomes A are only probable rather than invariable effects of their reference-property causes R. If the subjunctive conditional is embellished with a sign of causal strength u ('$p =u\Rightarrow q$') when the relationship is deterministic (or of universal strength u) and n ('$p =n\Rightarrow q$') – where n belongs to the inclusive [0,1] interval – when the relationship is indeterministic (or of probabilistic strength n), then these different types of causal conditional can be easily distinguished [Fetzer (1981)].

The use of these signs provides the foundation for a syntax that is sufficient for distinguishing assertions of any of these kinds from the others. Note in particular that deterministic causal conditionals entail corresponding subjunctives, but not conversely, while probabilistic causal conditionals do not:

(XVII) (4) '$p =\mathrm{u}\Rightarrow q$' entails '$p \Rightarrow q$', so
(5) '$p =\mathrm{u}\Rightarrow q$' entails '$p \rightarrow q$'; but
(6) '$p =\mathrm{n}\Rightarrow q$' does not entail '$p \Rightarrow q$', and
(7) '$p =\mathrm{n}\Rightarrow q$' does not entail '$p \rightarrow q$', either.

The ontology that underlies this account, therefore, remains the same as that discussed in Chapter 2, when we distinguished between deterministic and indeterministic as distinct kinds of causal systems. For deterministic systems are those for which, given the same input, the same output invariably occurs (without exception), while indeterministic systems are those for which, given the same input, one or another output within the same class of outputs (say, *O*1, *O*2, ..., *Om*) invariably occurs (without exception) with some probability. [The various forms of lawlike sentences of these kinds is displayed by (XX).]

That distinction, however, was drawn between deterministic and indeterminstic causal systems as two kinds of closed causal systems, where systems are *closed* when their reference-property descriptions take into account the presence or the absence of every property whose presence or absence makes a difference to (the strength of the causal tendency to produce) such an outcome. The properties whose presence or absence makes such a difference to an outcome, moreover, are known as "causally relevant" (or, more generally, "nomically relevant") properties, where nomic and causal relevance need to be clearly distinguished from mere "statistical relevance". The differences that are involved here are enormously important because, on the one hand, relations of nomic relevance can be employed to *explain* relations of statistical relevance, and, on the other, relations of statistical relevance can be employed as *evidence* in drawing conclusions concerning relations of nomic relevance. [Cf., Hempel (1968), Salmon (1971), and Fetzer (1981), esp. Part II.]

MAXIMAL SPECIFICITY

Statistical relevance is a frequency-based notion that concerns relative frequencies, while causal relevance is a propensity-based notion that concerns strengths of tendencies. They thus reflect different conceptions of probability. Their differences are important for all kinds of arguments, including analogical reasoning. Consider the comparison of two automobiles of the same make, Z-Cars, let us say. Suppose we know that the first Z-Car can accelerate from 0-60 in 4.5 seconds and want to know whether or not the second will also. We observe that both of these Z-Cars are four-door, with stick shifts, leather upholstery, and 280-cubic-inch engines. These reference properties might support an inference. If the rate of acceleration of other Z-Cars has varied with these properties, then by taking all of those that have made a statistical difference in the past, it should be possible to draw an inference concerning, say, a frequency-based expectation that the second Z-Car can accelerate from 0-60 in 4.5 seconds, too. Those reference properties that make no difference to this expectation are thus irrelevant.

In the case of propensity-based expectations, by contrast, the inferential situation is more complex, since these statistical differences may be regarded as indirect evidence of the causal relations that directly determine how rapidly acceleration can take place. Suppose, for example, that purely as a function of past statistics, the color a car is painted makes a difference to the frequency-based expectation of the presence of this attibute. Since this reference property, under these conditions, qualifies as a *statisically relevant* property, there are no grounds for regarding it as evidentially irrelevant on the basis of the frequency criterion. There is no latitude here to disregard this property on theoretical grounds. The alternative account, however, invites theorizing about the causal properties whose presence or absence makes a difference to an attribute, where the color a car is painted could be disqualifed as not a *causally relevant* property in arriving at propensity-based expectations that depend upon a more theoretical approach.

These concepts of relevance can be made more formal, of course, using the notion of a set of reference predicates, "$F1$", "$F2$", ..., relative to some reference predicate, "R", and an attribute predicate, "A", where the notion of statistical relevance can be formally defined as follows: property F is *statistically relevant* to outcome A, relative to R, if and only if

$$\text{(XVIII)} \quad SF(A/R\&Fi) \neq SF(A/R\&\text{-}Fi).$$

This means that the statistical frequency SF for A, given R and Fi, is not equal to the statistical frequency SF for A, given R and not-Fi, relative to an appropriate sequence of trials. Exactly how numerous or how varied a sequence of trials must be to be appropriate, however, is a difficult problem in the theory of statistical inference [cf. Fetzer (1981), Ch. 8]. In a similar fashion, the notion of causal relevance can be formally defined as follows: property F is *causally relevant* to attribute A, relative to R, if and only if

$$\text{(XIX)} \quad [(R\&Fi) = m \Rightarrow A] \ \& \ [(R\&\text{-}Fi) = n \Rightarrow A],$$

where $m \neq n$. This means that the strength of the causal propensity for R and Fi to bring about A differs from that for R and not-Fi to bring about A. The relations between them are such that, when (XIX) is the case, probably (XVIII) is the case; and when (XVIII) is the case, it is likely that (XIX) is the case. But spelling all this out is not a simple matter [Fetzer (1981), Part III].

These conditions, moreover, can be employed to define two distinct notions of what it is to be a "closed system", one of which reflects relations of statistical relevance, the other relations of causal relevance. Since the first merely reflects a "statistical homogeneity" conception based upon relations of statistical relevance, however, it should be viewed as capturing a special kind of accidental property that might or might not be generated by probabilistic laws based upon causal-strength relations. Note, in particular, that relative frequencies are properties of populations collectively, while causal propensities are properties of their individual members. The occurrence of any relative frequency for an attribute *A* is compatible with the presence or absence of that property by specific individual members of such a population. But the occurrence of a causal propensity of a corresponding strength for an outcome *A* is a property that is possessed by each of the individual members of populations that possess that reference property.

When a sentence happens to be a law, we can neither discover nor create instances of its reference property that can be separated from its attribute, no matter how hard we try, without losing that reference property. To change the melting point of something gold, for example, it would be necessary to discover some process or procedure – such as bombarding its nucleus with protons – that would change its atomic number, because this melting point is a permanent property of everything that has that reference property. Moreover, if there were other properties whose presence or absence made a difference to the melting point of things that are gold – the atmospheric pressure, for example – those additional properties would-have to be taken into account in fashioning that sentence for it to be a law.

The truth conditions for lawlike sentences, therefore, are very strict, indeed; for, in technical language, their reference property descriptions have to be *maximally specific*, such that for every predicate that designates any property whose presence or absence makes a difference to the presence or absence of that attribute or outcome, either that predicate '*F*' or its negation '-*F*' must be included. This condition can be expressed formally as follows:

> *The Requirement of Maximal Specificity*: If '*p*' is a true lawlike sentence, '*K*' is the reference-property description of '*p*', and '*F*' is any predicate that is nomically relevant to the truth of '*p*', then either '*F*' or its negation '-*F*', but not both, must be entailed by the reference-property description '*K*'

[cf. Fetzer (1981), pp. 50-51]. For if a true generalization allows of any exceptions, then it cannot be a law; and if a lawlike generalization allows any exceptions, then it cannot be true. Mere relative frequencies are not laws.

Epistemic Scorekeeping. The requirement of maximal specificity seems to provide a framework for understanding important dimensions of ordinary conversations as well as of scientific investigations. When a reference property happens to be complex, then it consists of a combination of other properties that might be distinguished from one another. Or, to make the point another way, a reference predicate 'R' may be logically equivalent to some conjunction of reference predicates '$F1$', '$F2$', ..., 'Fm', relative to the attribute predicate 'A'. In such cases, lawlike sentences of the three forms we have considered – '$p \Rightarrow q$', '$p =u\Rightarrow q$', and '$p =n\Rightarrow q$' – could be formulated by means of an enhanced prediate calculus in these two different ways:

(XX) (1) (a) $(x)(t)(Rxt \Rightarrow Axt)$, or
(b) $(x)(t)[(F1xt \;\&\; F2xt \;\&\; \ldots \&\; Fmxt) \Rightarrow Axt]$;
(2) (a) $(x)(t)(Rxt =u\Rightarrow Axt^*)$, or
(b) $(x)(t)[(F1xt \;\&\; F2xt \;\&\; \ldots \&\; Fmxt) =u\Rightarrow Axt^*]$;
(3) (a) $(x)(t)(Rxt =n\Rightarrow Axt^*)$, or
(b) $(x)(t)[(F1xt \;\&\; F2xt \;\&\; \ldots \&\; Fmxt) =n\Rightarrow Axt^*]$;

where (l)(a) and (1)(b) are logically equivalent formulations of the "simple" form of lawlike sentences, while (2)(a) and (2)(b) and (3)(a) and (3)(b) are equivalent formulations of deterministic and indeterministic "causal" laws. To understand how these lawlike sentences ought to be understood, a sentence of form (1)(a), for example, would be read, 'For all x and all t, if x were an instance of property R at t, then x would be an instance of property A at t'. A sentence of form (2)(a), in turn, would be read, 'For all x and all t, x's being R at t would (invariably) bring about x's being A at t^*', i.e., with strength equal to u (where t^* is equal to or later than t). And a sentence of form (3)(a), by comparison, would be read, 'For all x and all t, x's being R at t would (probabilistically) bring about x's being A at t^*', i.e., with strength equal to n. All of the other forms would be read similarly.

In ordinary conversations, of course, we often do make use of crude generalizations and "rules of thumb", such as that if a match is struck, it will light. We know that there are many conditions under which this is not true, however, such as when the match is wet, not enough oxygen is present, the match is not struck the right way, etc. Assuming that all of this is *background knowledge* to the participants in a conversation, then a specific case of a match being struck could be qualified by background knowledge into a complex form of reasoning involving an antecedent of a causal conditional, for example, corresponding to each of those special conditions that must be satisfied for an outcome of that kind to actually occur. Each of those special conditions, in other words, could be covered by a reference predicate '*F*1', '*F*2', ..., through '*Fm*' as might be required.

Assume, for example, that the reference predicate in this case is logically equivalent to the conjunction of '*F*1' and '*F*2' and ... and '*Fm*'. Then the following two deterministic causal conditionals would be equivalent:

(XXI) (1) $(x)(t)[(F1xt \;\&\; F2xt \;\&\; \ldots \&\; Fmxt] =u\Rightarrow Axt^*]$, and
(2) $(x)(t)[(F1xt \Rightarrow (F2xt \Rightarrow (\ldots \Rightarrow (Fmxt =u\Rightarrow Axt^*))))]$.

Sentences of these specific forms would be read as follows: (1) 'For all x and all t, x's being $F1$ and $F2$ and ...and Fm at t would invariably bring about x's being A at t^*'; and (2) 'For all x and all t, if x were $F1$ at t then if x were $F2$ at t then if x were ... at t, x's being Fm at t would invariably bring about x's being A at t^*'. These are clearly just two different ways of saying the same thing, which shows that they are logically equivalent. [However, see Fetzer (1981), pp. 148-149.]

If we were to assume that the participants in a conversation in ordinary language were equally familiar with (or knowledgeable about) when matches can be expected to light, then, when they exchange remarks like

(XXII) "If this match is struck, it will light",

they must do so in relation to their respective belief sets not only about matches in general but also about this match in particular, as aspects of the contexts in which they find themselves. If one of them were to make the remark (possibly as a causal consequence of a perceptual inference),

(XXIII) "But it's wet!",

no doubt they would be inclined to reassess their attitude toward (XXII), perhaps revising their beliefs on the basis of this new evidence, to assert:

(XXIV) "If this match is struck, it might not light", or
"If this match is struck, it will not light", or
"This match probably won't light", etc.

Thus, the process of keeping track of the relevant assumptions that are being made concerning these various exchanges could be well described as a matter of "conversational scorekeeping", in the suggestive phrase of David Lewis (1979). [Lewis' approach has been applied to discourse that involves conditionals by Nute (1980), within a more general framework.]

Similar conditions also obtain for scientific investigations generally in attempting to explain or in attempting to predict the occurrence of particular events. What this means, therefore, is that there appears to be a perfectly general pattern of reasoning that might be employed to establish a framework for representing the logical structure of discourse in ordinary and in scientific conversations and investigations. Insofar as this pattern reflects the beliefs of the participants in these conversations, it might be referred to as "epistemic scorekeeping", where the satisfaction of appropriate antecedents, relative to background beliefs, serves as evidence in relation to the assertion of the consequents of conditionals of these kinds. And there appear to be past-oriented and future-oriented species of each.

Thus, when an attribute has already occurred, t^* will belong to the conversational past, and the discovery of antecedent conditions at some prior time t whose satisfaction would justify inferring 'Axt^*' will have the character of an explanation. When the attribute has not yet occurred, t^* will belong to the conversational future, however, and the discovery of antecedent conditions at some prior t whose satisfaction would justify inferring 'Axt^*' will have the character of a prediction. The occurrence of explanatory and of predictive patterns of reasoning, therefore, is not limited to scientific investigations but commonly takes place in ordinary conversations, where the difference between them tends to be a matter of degree with respect to the extent to which the role of the requirement of maximal specificity is (more or less) explicitly acknowledged and of the extent to which efforts are expended to systematically isolate and identify the factors that are nomically relevant thereto [Fetzer (1981), p. xii].

The Frame Problem. The inferential situation is somewhat more complicated than this, if only because predictions can be made on the basis of statistical knowledge of relative frequencies in the past as well as on the basis of inductive knowledge of natural laws. The benefit of inferences drawn on the basis of natural laws, however, is that laws as properties of the world cannot be changed and cannot be violated: laws that have obtained in the past will obtain in the future, necessarily, as functions of the permanent property relations that they embody. The hazard with inferences drawn on the basis of relative frequencies, therefore, is that relative frequencies as properties of the world can be violated and can be changed: frequencies that have obtained in the past need not obtain in the future, necessarily, as functions of the transient property relations that they may represent. We can refer to this distinction as *the predictive primacy of scientific knowledge* over other beliefs about the future.

Thus, arguments in which lawlike sentences occur as premises tend to possess both explanatory and predictive significance, while those in which frequency sentences occur instead can have predictive, but not explanatory, utility. The conditions that must be satisfied for lawlike sentences to serve as a foundation for predictive and for explanatory arguments are perfectly general. They may be formulated as follows, where an "explanation" consists of an *explanans* and an *explanandum:*

(XXV) A set of sentences *S*, known as "the explanans", provides an adequate, nomically significant, scientific explanation for the occurrence of a singular event described by another sentence *E*, known as "its explanandum", in relation to some language, **L**, if and only if:

(1) the explanandum is a deductive or a probabilistic consequence of its explanans;
(2) the explanans contains at least one lawlike sentence of simple or of causal form that is actually required for the deductive or probabilistic derivation of the explanandum from its explanans;
(3) the explanation satisfies the requirement of strict maximal specificity with respect to its lawlike premises; and
(4) the sentences constituting the explanation – both the explanans and its explanandum – are true, relative to language **L**.

The requirement of strict maximal specificity (RSMS) ensures that the reference-property description of the lawlike premises that occur in such an explanation must include *all and only* those predicates that are nomically (or causally) relevant to the occurrence of the explanandum phenomenon [cf. Fetzer (1981), Part II; Salmon (1984); and Fetzer (1987) for elaboration].

These conditions support the conclusion that, from the logical point of view, there are two kinds of causal explanation, depending upon whether the general law(s) invoked in the explanans are deterministic or indeterministic in kind. If the laws invoked in the explanans are deterministic, then the logical properties of the relationship between the sentences constituting the explanans and its explanandum will be those of (complete) deductive entailment. If they are indeterministic, however, these logical properties will be those of (partial) deductive entailment. Consequently, these two kinds of explanation can be called "deterministic-deductive" and "indeterministic-probabilistic", respectively.

Perhaps most important from the perspective of artificial intelligence is that epistemic scorekeeping suggests a solution to the frame problem, which Charniak and McDermott have recently characterized as follows:

> The need to infer explicitly that a state will not change across time is called the frame problem. It is a problem because almost all states fail to change during an event, and in practical systems there will be an enormous number of them, which it is impractical to deal with explicitly. This large set forms a "frame" within which a small number of changes occur, hence the phrase. [Charniak and McDermott (1984), p. 418]

The problem, in other words, is that of ascertaining which states change and which do not change during some causal sequence. By relying upon epistemic scorekeeping, especially in relation to scientific knowledge, it ought to be possible, in principle, to cope with the difficulty that is posed. [For recent work on this issue, see Pylyshyn (1986) and Brown (1987).]

There appear to be two distinct aspects to the frame problem that deserve to be distinguished, where some phenomenon, attribute, or outcome of interest has been specified in relation to some corresponding reference property. We might want to know if this will light, if Zap can fly, and the like, where the thing referred to by "this" is a match, the object named by "Zap" is a bird, etc. The first aspect concerns whether or not the system under consideration qualifies as an *open* or as a *closed* system, relative to the phenomena, attributes, or outcomes of interest. The second concerns whether or not the system under consideration, when it happens to be a closed system, requires a *finite* or an *infinite* set of predicates for its complete description. For, as we are about to discover, theoretically adequate solutions for frame problems only appear to be possible, in principle, when closed systems can be appropriately described by finite sets of predicates.

Both aspects revolve about the extent to which the requirement of maximal specificity has been satisfied by means of a specific reference-property description. If the object of interest is a wooden match that is not wet and is being struck in a specific manner, then whether or not a certain outcome, such as lighting, will occur still depends upon the presence or the absence of those other properties, such as oxygen, whose presence or absence makes a difference to the occurrence of that outcome. When the presence or the absence of every relevant property with respect to a specific outcome has been taken into account by means of a maximally specific description, that system qualifies as a closed system, but otherwise not [cf. the definitions introduced in Chapter 2]. In the case of a closed system, it is possible to predict – invariably or probabilistically – precisely how that system will behave over an interval of time $t^* - t$ when those properties are instantiated at time t and the outcome occurs at t^*, so long as the laws of systems of that kind are known.

When either (a) the laws of systems of that kind are not known or (b) the description available for that system is not closed, however, then precisely how that system would behave over a corresponding interval of time $t^* - t$ cannot be predicted with – invariable or probabilistic – confidence, because essential information remains unknown. Even when the laws of systems of that kind are known, moreover, that knowledge could be incomplete when there is no end to the number of factors whose presence or absence makes a difference to the occurrence of the outcome of interest. An example that might plausibly fall within this category, moreover, might be that of human behavior, insofar as there seems to be no end to the variety of motives and beliefs that could make a difference to the behavior that causal systems of this kind display, a problem to which we shall return in Chapters 7 and 9.

In their depiction of the frame problem, Charniak and McDermott tend to blur some of these distinctions in at least two different ways. For they maintain, in the first place, that almost all states fail to change during an event, which might be true for some systems that are closed but is untrue for those that are open. Even in the case of a system that is closed, determining which properties will change and which will remain unchanged depends upon the kind of outcome taking place: during a nuclear explosion, for example, a lot of states tend to change. And they also maintain, in the second place, that the frame problem cannot be solved when the number of relevant causal factors is high, which has the unfortunate effect of pitting truth against practicality. There can be tension here, without any doubt; but it appears to be a tractable problem so long as these factors are finite.

In both of these cases, Charniak and McDermott seem to provide *a priori* answers to synthetic questions. We cannot know the percentage of states that are subject to change during an event apart from experience, because our knowledge of the course of events is synthetic knowledge. And we cannot know that exceeding some fixed number of finite factors has to render this problem unsolvable without taking into account the capabilities of our machines. It is always possible, of course, that the number of factors that are relevant to an outcome exceeds the capacity of currently available machines. But development of successive generations of computing machines has brought with it enhanced abilities to deal with increasing numbers of factors, which tends to mitigate this difficulty for finite sets of predicates.

It thus appears as though the frame problem can be solved in principle, at least, for closed systems involving only finite sets of relevant properties. Whether or not it can be solved in practice, of course, depends on the capabilities of the available machinery, which in turn depends upon the current state of technology (so long as matters of computational complexity do not prove intractable)! The function **cond** in the programming language Lisp provides a potential procedure for implementing causal conditionals whenever their finite, maximally specific antecedents are known. The function,

```
(F1)  (cond (exp11 exp 12 exp13 ...)
            (exp21 exp 22 exp23 ...)
            (exp31 exp 32 exp33 ...)
               .
               .
               .
            (exp n1 exp n2 exp n3 ...))
```

for example, provides an excellent illustration [Wilensky (1984), p. 55]. For in evaluating a **cond**, Lisp examines the first **cond** clause, where the entire list **(exp11 exp12 exp13 ...)** qualifies as one clause. If the first element of this clause, **exp11**, is true (that is, not-**nil**), Lisp will continue down that clause, evaluating **exp12, exp13, ...**until it comes to the end. If they are all true, Lisp returns the last element of that clause as the value of the **cond**. Otherwise, it ignores the other elements of that clause and proceeds to evaluate the next clause and the next, until eventually either some **cond** is satisfied or none of them is, in which case Lisp will return the value **nil**.

With respect to a finite set of predicates, "*F*1", "*F*2", ..., "*Fm*", relative to a specific outcome predicate, "*A*", the Lisp implementation might resemble:

```
(F2)  (cond (F1xt F2xt ... Axt*)
            (F1xt -F2xt ... -Axt*)
            (-F1xt F2xt ... Axt*)
                .
                .
                .
            (-F1xt -F2xt ... -Axt*))
```

where each of these clauses represents the relevant antecedents of one such maximally specific reference-property description and their last elements indicate the outcome that will occur when all of those antecedents happen to be satisfied. While this specific function could be straightforwardly adapted for subjunctive and deterministic causal conditionals, an enhancement could be designed for indeterministic causal conditionals as well, where the values returned might be the strength of the tendencies for such outcomes to occur. These procedures are not enough to cope with other types of cases in which the tensions that can arise between truth and practicality are not so easy to overcome, however, creating problems which we shall review in Chapter 6.

Before concluding this discussion, it might be worth mentioning that the semantics appropriate for subjunctive and causal conditionals is a form of *maximal change* semantics rather than the kind of *minimal change* semantics proposed by Robert Stalnaker (1968) or by David Lewis (1973). Thus, while their semantics depend upon assuming that possible worlds that are other than the actual world are as much like the actual world as possible, the semantics assumed here permits possible worlds to differ from the actual world in all respects except those specified by the maximally specific reference-property descriptions and certain dispositional properties that attend them. For such an approach, the only case in which the truth value of these stronger conditionals is determined by those of the corresponding material conditionals is when the latter are false. [For discussion of these types of model theory, see Nute (1975) and Rankin (1989); on the theory appropriate to the present account, cf. Fetzer and Nute (1979) and (1980).]

6. EXPERT SYSTEMS

Perhaps the most important conception that the AI community has yet devised is that of an "expert system". This involves the implementation of a body of expert knowledge in the form of a computer program in order to make that knowledge available to anyone who has the capability to utilize that program. If we define a *domain* as a class of problems that stand in need of solutions and an *expert* as someone who is prepared to solve them, then an *expert system* can be characterized as a program that, when suitably employed, can transmit an expert's solutions to those confronted by problems within that domain. The appealing feature of this conception is that limited expert knowledge can be provided in almost unlimited supply for those who stand in need, thus rendering ordinary non-experts potentially as knowledgeable as the most extraordinary experts in domains as diverse as business, medicine, and psychology [see Hayes-Roth, Waterman, and Lenat (1983), Buchanan and Shortliffe (1984), and Waterman (1986)].

The enterprise of constructing an expert system, moreover, involves at least three stages, which are governed by the aim, objective, or goal which that system is intended to achieve. Suppose, for example, that a potential user of such a system wanted to have an expert system for the diagnosis of disease. The first stage in the construction of such a system would involve the specification of the specific diseases to be diagnosed and selection of an expert with respect to that domain. The second stage would involve acquiring the knowledge that is required from the expert selected and expressing it in a form suitable for translation into a program. The third stage would involve taking that information and implementing it by means of a programming language into a suitable (usable) program.

Provided that a suitable problem domain and a suitable expert have been selected, those who perform the processes of acquiring the relevant knowledge from a designated source and preparing it for use by the programmer are known as "knowledge engineers". Those who take the information prepared by the knowledge engineers and translate it into a usable program, of course, are "software engineers". But the success of the endeavor hinges upon the expertise of the expert selected, since even the world's best knowledge engineers in collaboration with the world's best software engineers cannot overcome the deficiencies and limitations that can result from making a wrong choice with respect to those whose expertise provides the source for the knowledge on which the system is based.

THE GENERAL CONCEPTION

Although the benefits that can be derived from the utilization of an expert system are enormously appealing on their face, it might be worth reflecting that many of the same features cited above are also available to the public through the publication of articles and books, whereby a specific expert's expertise can be made available to any member of the community at almost any time and place, so long as he has the ability to read. But if an expert's expertise can be stored in and retrieved from books and articles as well as computer programs, it may seem difficult to justify the expenditure of large sums on hardware and software in lieu of the allocation of small sums for articles and books. This fiscal comparison, of course, requires its own qualifications, since the computer side of the ledger includes production costs, while the library side does not. If the expenses incurred in purchasing printing presses, in typesetting, galley-proofing, and the like, are taken into account, the numbers might change but not the balance. If expert systems are better than libraries, the reason should be explained.

Assuming the availability of programs and computers, on the one hand, or of books and libraries, on the other, one advantage of expert systems appears to be their convenience of utilization, provided, of course, that appropriate personnel have the appropriate training. The information stored in a program can be readily retrieved, under suitable conditions, by an operator sitting at a fixed location, whereas the information stored in a library is less readily retrieved, under suitable conditions, by a reader moving to a book's location. A well-constructed program, moreover, facilitates the rapid accessibility of specific information by means of electromagnetic capabilities that greatly exceed the speeds with which most readers are prepared to derive most information from most books, even when they are provided with a detailed index. Whenever time is as precious as or more precious than money, therefore, expert systems would appear to have the rather substantial edge.

Another important advantage, no doubt, is that computer programs can be specifically designed to handle special classes of problems for specific individual consumers, where those programs can perform tasks that otherwise would require the employment of human minds. Indeed, human history has witnessed at least three transformations in the distribution of human effort that properly qualify as "revolutions": (i) the Agricultural Revolution, which put nature to work for man in raising crops and cultivating gardens; (ii) the Industrial Revolution, which put machines to work for man to ease his physical labor; and (iii) the Computer Revolution, which put machines to work for man to ease his mental labor. And the potential for ordinary problem-solving programs (such as pocket calculators) as well as for expert problem-solving programs (such as diagnostic systems) is difficult to overemphasize – at least, so long as we do not ask of them tasks that they are unable to perform.

The most significant difference between libraries and computer programs emerges at this juncture, for it would be a mistake to imagine that the machines are nothing more than electronic counterparts of libraries as electronic information and retrieval systems. The interactive capabilities available to the users of these programs are completely superior to those available to the readers of those books, insofar as computers and programs, when properly designed, function as active problem-solvers rather than as passive information-providers. Computers and programs possess the ability to actively participate in solving problems by making their problem-solving expertise available to their system's users by means of proposing answers and asking questions in response to inquiries falling within their domains of expertise. However stimulating a book may be to an avid reader, the mental activities brought about through reading are those of the reader alone, while users of computers and programs secure the added advantage of engaging the problem-solving abilities of that system by means of question-answer exchange.

Perhaps the most fascinating aspect of the expert-systems concept, therefore, appears to be the idea of making every user of a computer program potentially as knowledgeable as the expert upon whose knowledge that system has been built. This idea harmonizes well with the political ideals of an egalitarian society, not to mention the profitability aspirations of a capitalistic economy. Thus, while the expert-systems concept naturally presumes the existence of "experts" who know more about the solutions to a certain class of problems than others do, it extends the promise of making that expertise available to all who have the more rudimentary skills required to process inputs and to interpret outputs, enhancing the prospect of elevating the problem-solving abilities of the community as a whole. To the extent to which the contents of an expert's mind can be placed at the disposal of a program's user, it would seem, expert systems promise the virtues of mind transplants without the vices of brain surgery. But they have limitations, nevertheless.

The Choice of an Expert. Since "knowledge" itself is a term of success that implies the attainment of something worth having, it might be important to return to the procedures that lie at the core of the conception of expert systems as knowledge-based systems. The crucial aspects of the expert system process turn out to be the acquisition, the representation, and the utilization of the expert's expertise. A knowledge-engineer has to acquire knowledge from the expert by some (perhaps quite complex) appropriate means, typically involving question-and-answer sessions, which may require a great deal of time to secure the relevant information. A software-engineer then has to represent the information that has thereby been obtained in a form that is suitable for the machine's language and for the user's language in the course of composing an appropriate program, often by means of production rules that assume the form of conditionals. And a user's manual has to be devised to make the information reflected by that program relatively easy to derive for operators who may not be experts with respect to this domain.

Still more important than the acquisition, representation, and utilization problems that expert systems generate, however, is the choice of the experts themselves. This turns out to be a greater problem than might be supposed, insofar as, apart from the more elementary information that seems to be the common knowledge of beginning students within a given domain, advanced research on difficult subjects has the propensity to spawn conflicting points of view. Thus, while the leading figures within various disciplines almost always agree upon their more elementary facets, this is not the kind of knowledge that qualifies as "expertise". And when it comes to complicated problems that demand more specialized knowledge, "the experts" almost always disagree. Without experts there can be no expert systems, however, because an "expert" has to be selected (or created) in order to build an expert system.

The comparison with libraries and books tends to be instructive here, for no one would suppose that it is possible to have something to read without the necessity of choosing a book – or of having one chosen! The very same situation obtains with respect to computers and programs, since it is similarly impossible to build an expert system without choosing an expert – or having one chosen! The means whereby such a choice is made, therefore, should be viewed as the crucial question underlying the construction of any expert system, though it does not invariably have to be a matter of great moment. An expert system for handicapping horses could not be devised without relying upon some expert, but choosing, say, Jimmy the Greek might not have to qualify as especially important. An expert system for conducting nuclear warfare, by comparison, could not be constructed without relying upon some expert, but the selection of Dr. Strangelove would be an ominous choice. The signficance of the choice thus varies with the importance of its consequences.

Consider three among a large number of alternative methods of selecting some expert for the construction of an expert system within such a domain. One mode of choice would be to make an *arbitrary* selection from the members of the general population, which – apart from very special cases – would result in a worthless candidate, unless the questions before us happened to concern, say, the meaning of life, a matter about which every man may be his own expert. Otherwise, of course, arbitrary choices have obvious defects. Yet a second mode of choice would be to make a *sociological* selection, relative to the opinions of researchers within that specific domain. No doubt, this mode tends to be the preferred method of choice among current system builders, since it offers a relatively uncomplicated and a reasonably practical solution to a troublesome problem. A third mode of choice, however, would involve making a *methodological* selection relative to the reliability, the efficiency, or the effectiveness of various epistemic procedures within that same domain.

The Nature of Expertise. Although the deficiencies of arbitrary choices are easy to discern, the defects that attend sociological choices are far less obvious but are not therefore any less serious. One difficulty, of course, is that popularity (even among experts) is no guarantee of knowledge, competence, or truth, a phenomenon familiar to politicians but whose broad significance has not always been acknowledged. Even research communities tend to adopt standards of their own, especially in relation to their accepted "paradigms", which serve to exclude others whose approach differs from theirs, either on grounds of ignorance, or of false belief, or even of incompetence. The situation here, however, is a double-edged sword, for surely membership in a research community cannot in and of itself establish the truth of beliefs, the knowledge or the competence of its members. The American Astrological Association, the Crystal-Ball Gazers of America, and the Tea Leaf Readers Society might all have members in good standing, but the potential benefits of their presumed expertise still remains an open question.

Indeed, the situation with respect to the sciences themselves is really no different, in principle, insofar as the possibility of quackery or of malpractice cannot be ruled out. Not only do research communities tend to adopt their exclusive paradigms, but there appears to be no suitable foundation for choosing one group of experts as bona fide or for regarding some other as fraudulent unless recourse is ultimately made to methodological principles. The expertise that an expert possesses, after all, consists of at least two distinctly different capacities and abilities: first, an expert possesses certain sets of beliefs that qualify as the *knowledge content* appropriate to his domain of expertise; second, an expert also possesses certain talents and skills that qualify as the *epistemic abilities* appropriate to that same domain. One of the differences between an expert and a fraud tends to be that a fraud often appears to possess an expert's knowledge content but cannot support knowledge claims because he lacks the epistemic skills.

Scientific Expertise. This problem invites consideration of the differences between two accounts of the nature of scientific expertise, which reflect the distinction between sociological and methodological criteria for an expert's selection. There are slender grounds for believing that the scientific theories we happen to accept today are destined to endure a different fate than their predecessors endured. What counts as "scientific knowledge" for a certain time and place, therefore, turns out to be a function of the set of beliefs that the members of some scientific community Z happen to accept at time t, where these beliefs satisfy conditions of rational belief like (CR-1) to (CR-4) without assuming that these beliefs are always going to turn out to be true. The scientific knowledge of a community Z at a time t thus consists of what Karl Popper (1965) would describe as the best guesses that have withstood our best attempts to overthrow them, where Popper especially appreciates the importance of testing lawlike sentences by trying to show that they are false. Since laws of nature assert the nonexistence of processes or procedures that would permit the separation of properties of different kinds, the best way to test them is to attempt to establish that these exist, where the failure of efforts to refute laws counts as evidence supporting their truth.

A different aspect of scientific knowledge has been emphasized by T. S. Kuhn (1970), who suggests that the scientific knowledge of a time consists of the beliefs that are accepted by the scientific community of that period. These views appear to be quite difference on their face, since Popper seeks to emphasize the importance of methodology in testing lawlike hypotheses, while Kuhn attempts to stress the importance of consensus in reaching community decisions. Kuhn takes for granted that the members of a scientific community possess the background, training, and other qualifications that are appropriate to that position. But the question then becomes one of determining what background, training, and qualifications are the right ones.

Whether the distance between their views is very great or very small, therefore, depends upon the kind of background, training and other qualifications presupposed for membership within Kuhnian scientific communiities. And the problem here becomes the abundance of alternative conceptions, which depend upon and vary with the methodological principles that are supposed to be appropriate for scientific inquiries. There are a number of competing alternatives, of which I shall characterize at least these three:

INDUCTIVISM:	DEDUCTIVISM:	RETRODUCTIVISM:
Observation	Conjecture	Puzzlement
Classification	Derivation	Speculation
Generalization	Experimentation	Adaptation
Prediction	Elimination	Explanation

Fig. 17. Alternative Conceptions of Scientific Methology.

Thus, "The Bayesian Way" with its reliance upon Bayes' theorem represents yet another alternative, where each one supports specific rules of reasoning.

The strengths and weaknesses of these positions are important to epistemology in establishing the foundations of scientific knowledge. The rules of inference that Inductivism supports, including the straight rule discussed in Chapter 1, for example, are restricted to the relative frequencies that can be observed during the course of the world's history and are not strong enough to warrant the discovery of natural laws. The rules of inference that Deductivism supports, including *Modus Ponens* and *Modus Tollens*, are purely deductive and cannot support the forms of ampliative inference that are essential to warrant the discovery of natural laws. When Deductivism embraces the positive significance of unsuccessful attempts to falsify theories as well as the negative significance of successful attempts to falisify them, however, it shades into Retroductivism, which envisions the discovery of natural laws as a form of abduction or "inference to the best explanation" [Fetzer (1983b)].

The study of methods of inquiry, however, falls primarily within the province of philosophers of science rather than of scientists themselves, principally because these issues are *normative* rather than *descriptive*. Discovering which principles and procedures are the most efficient, reliable, and effective involves different backgrounds, abilities, and skills than attempting to discover laws of nature themselves. As a consequence, even successful scientists may find it difficult or troublesome to articulate those principles and procedures that underlie scientific inquiries. And it should come as no surprise that the doctrines that are preached in the classroom are often not the same as the practices that are employed in the laboratory. It is not difficult to discover scientists who espouse Inductivism but whose work could never have been achieved by relying upon its standards [cf. Fetzer (1984)].

The problems that arise in selecting "experts" to build expert programs, therefore, are not easy to settle by sociological criteria, unless they happen to correspond with suitable methodologies. Unless the background, training, and skills possessed are those appropriate to scientific inquiries, even those with advanced degrees and professional affiliations may fail to possess the qualifications that are appropriate to their disciplines. Even the existence of a community adhering to established practices is not enough to satisfy this desideratum. For alchemy, astrology, and witchcraft are enduring practices whose practitioners may fail to carry them out but therefore qualify as *sciences* no more than do mortuary science, library science, and Christian science merely because they are so-called. The existence of quacks, charlatans, and frauds cannot be ruled out on the basis of mere numbers alone.

The bottom line appears to be that science cannot properly be viewed as what the members of a presumptive "community of scientists" do, since the qualifications for being a scientist require the capacity to pursue science by employing methods appropriate to that domain. The principles and procedures by means of which a discipline is defined, therefore, cannot be ascertained on the basis of empirical evidence alone. The Kuhnian conception of a community of scientists can be reconciled with the Popperian analysis of scientific method by making Popperian practice a condition for membership within a Kuhnian commuity. More generally, the conception of any discipline as a community of practitioners can be reconciled with the normative standards suitable to that discipline by insisting that the members of such a discipline are those whose own practice implements those specific standards.

EXPERT PRODUCTION SYSTEMS

Indeed, the problems that are involved in identifying suitable "experts" whose "expertise" might be worth knowing has been acknowledged by the expert-systems community, even with respect to "the standard conception" of systems of this kind. Donald Waterman, for example, has suggested that the necessary requirements for expert-system development concerning the performance of specific tasks within a special domain tend to include these:

(1) that the task does not require common sense;
(2) that the task requires only cognitive skills;
(3) that experts can articulate their methods;
(4) that there really do exist genuine experts;
(5) that these experts agree on these matters;
(6) that the task is not excessively difficult; and,
(7) that the task is not inadequately understood;

[cf. Waterman (1986), p. 129]. When cognitive skills involving the utilization of systems of symbols are distinguished from perceptual abilities and from practical skills, these requirements make a great deal of sense, especially in view of the crucial if implicit role of the requirement of maximal specificity.

Even though the expert-systems community does not appear to be preoccupied with the normative as opposed to the descriptive aspects of issues of this kind, these guidelines appear to be appropriate. The existence of inarticulate experts, no doubt, is almost as bad as the non-existence of experts, from the point of view of system development, since knowledge engineers cannot transcribe expertise that experts cannot articulate. And when the experts disagree, there is no logically consistent set of beliefs to program, where the content thereby represented might possibly be true. Moreover, when a task is not well understood, it becomes exceedingly problematic if not completely impossible to satisfy the requirement of maximal specificity, in the absence of which systems under consideration have to be "open" rather than "closed". And when there are more than a finite number of relevant variables, a task becomes too difficult to program for a finite machine.

Waterman has also suggested that the effort and expense that are involved in the construction of an expert system tend to be justified under some fairly specific kinds of circumstance, which include especially the following:

(a) when the task solution has a high payoff,
(b) when human expertise is being lost,
(c) when human expertise is scarce,
(d) when expertise is widely needed, or
(e) when expertise is needed in hostile evironments;

[cf. Waterman (1986), p. 130]. What is most striking about conditions (a) to (e) as opposed to (l) through (7), no doubt, is that (a) to (e) reflect historical and pragmatic states of affairs involving the relative distribution and relative importance of special kinds of expert knowledge as a function of a possible user's specific needs, which are independent of the state of knowledge within that domain. (l) through (7) reflect the programmability of a special knowledge domain, while (a) to (e) concern whether it will be programmed.

Notice, for example, that even when human expertise is being lost, unless there is a market for a corresponding expert system, there is little likelihood that one will be developed. The law of supply and demand operates here as elsewhere within the commercial marketplace, with all of the attendant inequities and inequalities that economic forces tend to bring about. If a wealthy businessman wanted an expert system to assist his wife with her makeup, it is not hard to imagine that a system of this kind could be designed for an appropriate fee. Yet even if some literary figure's techniques of criticism could be rendered in a form that would be suitable for programming, lacking a suitable funding source, the loss of that expertise becomes practically inevitable. The kinds of "payoffs" that matter here, alas, tend to be monetary or even military rather than intellectual or cultural, so that when they happily coincide, the result appears almost accidental rather than deliberate in kind.

Logical Structure. The knowledge that is acquired from an expert by the knowledge engineer is typically cast into the form of a "production system". Perhaps the most important feature of production systems from the logical point of view, moreover, is that they characteristically consist of sets of sentences having the form of 'if ___ then ...' conditionals. The following rules, for example, are adapted from a knowledge-based wine selection program:

Rule-1: If the meal has sauce and the sauce is spicy, then the best body is full.
Rule-8: If the meal's main component is fish, then the best color is white.

Fig. 18. Some Production Rules.

Other production rules incorporate what are known as certainty factors (CFs):

Rule-2: If the meal is delicate in taste, then (CF = 80) the best body is light.
Rule-6: If meat is served and it includes veal, then (CF = 90) the best color is red.

Fig. 19. Other Production Rules.

The purpose of this expert system, naturally, is to provide advice about the best wine to serve with a meal, depending upon the meal's principal ingredients, such as the type of meat, the kind of sauce, etc., leading to a selection [cf. especially ⟨M1⟩Wine, Version 1.1, Copyright (c) Teknowledge Inc. 1984].

Three features of production systems of this kind are most important for our purposes. The first is that this approach tacitly reflects specific commitments to particular *relevant properties*, which emerge from the decisions it renders about which factors are relevant to the selection of a wine to serve with a meal. The factors of *type of meat* and *kind of sauce* are properties whose presence or absence makes a difference to the selection that will be made. Such factors as *dining-room decor* and *price of wine* are properties whose presence or absence does not make any difference to the selection that will be made. Even if it appears as though other factors than those relied upon by this system should be taken into account, the specific features that are taken into account here evidently reflect either the expert's choice or the programmer's choice in constructing this specific individual system.

The second is that the 'if ___ then ...' conditionals that this system employs characteristically have antecedents ("the meal has sauce", "the meal's main component is fish", etc.) and consequents ("the best body is full", "the best color is white", etc.) that consist of singular sentences. This means that these specific rules do not involve quantification across conditionals, which means that they have a familiar logical structure, namely: that of sentences that can be adequately formalized by means of *sentential logic* [cf. Gustason and Ulrich (1973) and Quine (1951)], which includes rules like the following:

Modus Ponens (MP): If '$p \rightarrow q$' and 'p', infer 'q';
Modus Tollens (MT): If '$p \rightarrow q$' and '$-q$', infer '$-p$';
Transitivity (Trans): If '$p \rightarrow q$' and '$q \rightarrow r$', infer '$p \rightarrow r$'; etc.

Fig. 20. Some Rules of Sentential Logic.

where the letters 'p', 'q', 'r', ...are place-holders for unspecified sentences. These are generally representative of patterns of reasoning that can be executed by production systems, including chaining "forward" (from premises to conclusions) and "backward" (from conclusions to possible premises).

The third is that the occurrence of rules like those displayed by Figure 19 indicates that the logical structure of production systems cannot be completely identified with sentential logic but requires some form of supplementation by means of the incorporation of *certainty factors* (CF's), which might be based upon or related to subjective probabilities, relative frequencies, or strengths of causal tendencies. As examples of possible measures, consider:
first, confidence factors can be viewed as indicators of nomic expectability,

(NE) When '$Rxt =n\Rightarrow Axt^*$' and 'Rat' belong to a knowledge base *KB*, then the nomic expectability NE (or "CF") of 'Aat^*' in *KB* is n;

second, confidence factors can be viewed as indicators of relative frequency,

(RF) When '$Rxt \text{-}f\rightarrow Axt^*$' and '$Rat$' belong to the knowledge base *KB*, then the relative frequency RF (or "CF") of 'Aat^*' in *KB* is f

[cf. Fetzer (1983a)]. As a rule, objective measures are preferable to subjective measures, since they reflect forms of empirical knowledge as opposed to personal opinion, where system builders must make the proper choice.

Epistemic Contents. It should be clear that rules like those represented by Figure 18 and by Figure 19 are crude generalizations or "rules of thumb", even if they are derived from an expert. That this must be the case results from the realization that the relevant properties focused upon by such rules leave out of consideration factors that ought to be taken into account if one were attempting to satisfy the conditions appropriate to scientific laws. The construction of this specific system, after all, (which Teknowledge designed as a teaching device rather than as a serious advisor) is based on special assumptions concerning the factors that are relevant to the selection of a wine to accompany a meal, where these include the meal's main component, whether it has a sauce, whether it is spicy, etc., in arriving at a recommendation as to the type of wine (chablis, sauvignon blanc, chardonnay, etc.) to choose.

An expert system of this sort could count as a scientific-knowledge-based system, therefore, only if the relevant factors on which it is based represent a complete set of relevant factors, in which case it might satisfy the requirement of maximal specificity. That this system fails to do so should be obvious on several grounds, since it not only ignores the factor of price but also omits factors of health. No doubt, there is scant chance that users of such a system would follow its advice and drink red wine, for example, were they aware that they were allergic to red wine (unless they wanted to induce an an allergic reaction). Even more surprising, this system completely omits a recommendation as to label or brand, which is the most valuable kind of information that a connoisseur can provide to the amateur [cf. Prial (1988)].

The point is not that this expert system should be expected to take allergies into account, but rather that this must be a system of "rules of thumb" that can be counted on to offer appropriate advice for some (for many, for most) of the decisions that might be made by relying upon it (as a function of the historical conditions under which they happen to occur), but clearly not for all. Indeed, the advice that it extends is not always appropriate and might even be completely wrong (even fatal) if it were followed blindly by those who might be inclined to use it. This problem, of course, is a matter of *context*, including the motives and beliefs that the user brings to the system in the process of its utilization. We may not expect fatalities from the advice derived from a system for selecting wine. But other expert systems (for the selection of targets by warships engaged in combat, like the USS Vincennes on patrol in the Persian Gulf) may turn out to be less forgiving, not because those systems do not function properly but due to the faulty role of context.

The construction of an expert system thus presupposes a decision about the kind of knowledge that system is supposed to represent. Notice, especially, that a choice between different measures of relevance lies just below the surface with respect to their development. For criteria of *statistical relevance* on the basis of relative frequencies can supply one kind of objective foundation for the assignment of confidence factors to alternative outcomes, but criteria for *causal relevance* on the basis of causal tendencies can supply another. Statistical relevance reasoning has the virtue of relatively straightforward access to the information that is required in order to build a system, while causal relevance reasoning has the virtue of the predictive primacy of scientific knowledge over other kinds of knowledge about nature (Chapter 5).

The differences between knowledge of these two kinds, however, may or may not become apparent to the knowledge engineer, the software engineer, or the product user, who depend upon experts while taking for granted their expertise. Indeed, unless those involved in the design, development, and utilization of systems of this kind were aware of the fundamental differences between knowledge of these different types, they would have no grounds to differentiate between them. Yet the choice between kinds of knowledge for the construction of expert systems can be a matter of life and death. Those involved in the expert-systems business, therefore, appear to confront some serious ethical issues in relation to the expertise employed by their systems.

But this raises yet another problem, namely: how can we ever decide if an expert system of any kind should be relied upon? Notice that this is not a question about the sources of knowledge that have been drawn upon during the course of its construction by the knowledge engineer or the software engineer in developing their product. The problem is that systems such as these can only be properly employed when the roles for which they have been designed are adequately understood. For when the "rules of thumb" that they employ as a basis for their recommendations are appreciated *in relation to the background assumptions and the contexts for their ultimate use*, then those who rely upon this system for advice will receive the advice they desire. But if the factors that make a difference to their users are not the same as the factors that make a difference to the systems, the potential for mistakes – possibly mistakes of the most serious kinds – is unrestrained.

What this means (and it will come as no surprise) is that an expert system built upon "rules of thumb" does not provide exceptionless generalizations about its domain of expertise. Perhaps for this reason, if for no other, the composition of a user's manual assumes a significance beyond its mere ability to inform a user how to extract information from the system. Those who design and develop expert systems clearly incur an obligation to make explicit precisely how the information that such a system utilizes should be properly understood. For, if there are boundary conditions or other factors whose presence or absence makes a difference to the advice that is presented by that program, then those conditions and factors ought to be explained to potential users of that system. Those who rely upon the advice provided by expert systems must be trained in the interpretation of the advice that they provide, especially when misinterpretation could have serious effects.

Truth vs. Practicality. We thus confront another aspect of the problems we have previously encountered in relation to the frame problem. For the reason why "rules of thumb" are not exceptionless generalizations can also be explained by observing that these "generalizations" require interpretation. Indeed, the role of confidence factors (CFs) as displayed by Figure 19 raises this issue explicitly. Consider, for example, Rule R-2: If the meal is delicate in taste, then (CF = 80) the best body is light. Does this mean that, in the expert's opinion, 8 times out of 10, a light wine tastes best? If this is his subjective evaluation, however, can it not change from time to time and from expert to expert? Or does it mean that, during a random survey of 100 taste testers, 80% preferred a light wine all of the time? But if this is the case, does that imply that 20% of the time this choice will be wrong?

Even more important than developing a justifiable theory of confidence factors, however, are two remaining difficulties that were not adequately addressed in the course of our discussion of the frame problem that arise from the nature of these machines themselves. Clearly, it can be the case that the number of nomic (or causal) factors that are relevant to the occurrence of events of some specific kind might be very large. While there are reasons to believe that there are never more than a finite number of kinds of properties that are nomically (or causally) relevant to a specific outcome phenomenon, there are no grounds for doubting that at least some of them may be amenable to quantitative variation, if not in every case, at least in many instances. When the finite memories and processing capacities of digital machines are explicitly considered, therefore, the issue of practicality assumes a new dimension. For an unlimited variation in lawlike phenomena would be impossible to program into a finite machine at any one time.

When either the number of relevant factors is infinite or finite but beyond the capacity of the current generation of machines, practicality and truth stand in conflict. The practical version of "the frame problem" thus emerges at this juncture, where the question no longer remains, "What is true?", but rather, "What shall we do?" When only part of what is known can be effectively utilized by an expert system (due to limitations of space and time in processing data), a decision must be made, where the features of *decisions* are unlike those of *inferences*, in general. For "decisions" are usually irrevocable and involve choosing between restricted alternatives, while "inferences" are ordinarily subject to revision and the possible alternatives are virtually unlimited [cf. Fetzer (1981), pp. 236-237]. And it is not obvious that there are optimal solutions to decison-making problems.

At least two kinds of information would appear to be relevant in arriving at a decision concerning the information that ought to be incorporated into an expert system, when limitations of space and time dictate that decisions have to be made. One of these reflects the relative frequency with which events of a certain (reference) kind have occurred in the past, while the other concerns the relative importance that would attach to events of a certain (attribute) kind were they to occur. The development of systems for the diagnosis and treatment of medical disease, for example, might well take into account the relative frequency with which various diseases tend to occur in forming expectations as to the most probable kinds of diseases that will need to be diagnosed. But, even when diseases are relatively uncommon or even rare, consideration might equally well be given to the seriousness of the consequences that attend contraction of those diseases.

The differences between expert systems for selecting wines and expert systems for selecting weapons are relevant to this problem. We have the tendency to discount the problem of choice of wine because, even though decisions of this kind are frequently required, the consequences of making the wrong decisions are typically quite mild. We have the tendency to inflate the problem of choice of weapon, however, because, even though a decision to drop an atomic bomb is comparatively rare, the consequences of making the wrong decision are devastating ones, indeed. The problems that decision-makers confront are not restricted to the practical problems of implementation, moreover, since their theoretical dimensions are clear.

The discussion of rationality in Chapter 1, for example, may have offered at least the hint that we confront subtle problems on almost every side. As an illustration, consider the policy of maximizing expected utility, which is frequently endorsed as the most appropriate decision-making policy. That we should accept the belief that maximizing expected utility *is* the most appropriate decision-making policy itself, of course, reflects the rationality of our beliefs, in turn, for if there were stronger reasons to doubt the truth of such a claim than there were reasons to adopt it, it ought to be rejected (or, at least, left in suspense) instead. All optimizing decision-policies of which this one is an instance, however, confront serious difficulties of their own, especially in obligating us to insure that, whatever our decisions may turn out to be, their expected utilities cannot be *less* than those available by adopting *any* alternative other than those that we select [cf. Michalos (1973)].

The other difficulty concerns the fallibility of scientific knowledge itself, which appears to be an inescapable consequence of reliance upon inductive procedures in the search for natural laws. For the "scientific knowledge" of our time – or of any other time – is only the currently accepted beliefs of a specific commuity of inquirers, who might, after all, be making a mistake. It is ironic to realize that this is less a problem for scientific inquiry than it is for expert-system builders, since the members of the scientific community can adjust their beliefs in response to the arrival of new evidence, whether in the form of new hypotheses and theories or in the form of new observations and experiments, perhaps generated by advances in technology. Once an expert-system product has been placed in the hands of its users, however, decisions are going to be made that depend on its performance, whether or not the information in its data-base is current or superseded.

This suggests that the knowledge engineer, the software engineer, and the ultimate consumer have to confront some potentially very serious decisions concerning what to include and what to exclude in cases in which it is impossible to do everything one might like. Indeed, the mere implementation of scientific knowledge in the form of a program itself already appears to generate a moral dilemma. For if only some of the knowledge that is relevant to a domain can be implemented by means of a program, which portion of that knowledge should it be? The knowledge which is likely to be relied upon most commonly may not have the most serious consequences, while the knowledge with the most serious consequences is likely to be the least commonly relied upon. When tensions result at this juncture, which knowledge should be included and which excluded?

VARIETIES OF EXPERT SYSTEMS

Other problems yet remain. If there can be a fraudulent expert, then there can be a fraudulent expert system. For an expert system to properly qualify as more than an electromagnetic storage and retrieval system, it ought to have the epistemic abilities as well as the knowledge content experts represent. These epistemic abilities, in fact, are more important than that knowledge content *per se*, insofar as they provide the capacity for a system to acquire and to process the information that becomes the expert's knowledge to begin with. With these epistemic abilities, in other words, it should be possible, under suitable conditions, not only to store old knowledge acquired in the past but also to acquire new knowledge in the future. Such a capability, moreover, ought to appeal to expert-system builders, since one of the shortcomings of expert systems as they are currently conceived tends to be the necessity to update their knowledge bases as new knowledge becomes available or as old experts change their minds.

It would be a mistake, however, to draw the conclusion that the relevant court of appeal is cognitive psychology. No one has demonstrated that the methods which humans (including our own experts) happen to employ to solve a specific class of problems are the most effective, the most efficient, or the most reliable to secure those specific goals. The problem is not that those approaches that happen to be employed by ordinary human beings to solve ordinary problems or by exraordinary experts to solve extraordinary problems could not be discovered by empirical research, under suitable conditions, but rather that the objective that would thereby be secured would still be insufficient to establish the efficacy, the efficiency, or the reliability of those methods themselves. The pressing need, in other words, appears to be for normative evaluations of methodological alternatives rather than for empirical investigations of procedures that have been employed in the past.

These reflections raise the possibility that methodological considerations might provide an avenue of escape from the problem of selecting an expert presently confronting the expert-system builder. If the choice of an expert is to be a rational act as opposed to an arbitrary selection, then either that expert is selected *without* the advice of other experts, which would appear to be an unjustifiable selection, or the expert is selected *with* the advice of other experts, which would appear to be a justifiable selection – so long as the selection of those *other* experts is amenable to justification itself. Thus, an escape route from either begging the question regarding an expert's expertise or generating an infinite regress of experts to certify the expertise of other experts surely ought to be welcomed with relief. For methodological considerations can establish that certain goals are reasonable goals because they are attainable; that certain methods are reasonable methods, because they are conducive to the attainment of reasonable goals; and that reasonable methods have led to the attainment of reasonable goals in certain classes of cases, where those cases happen to satisfy these conditions.

Among the most important elements of expertise from a methodological point of view, moreover, are those features that distinguish an expert's capability to acquire and to process knowledge within its domain, such as those we have considered in Chapter 4, namely: the experts' (a) semiotic abilities, (b) perceptual capabilities, (c) deductive powers, (d) inductive powers, and (e) powers of imagination and conjecture. To make possible building a system that would implement these epistemic resources in the form of a program, therefore, they must be reformulated to satisfy the limitations that have been discovered to be imposed by the machine. This result, obviously, reflects differences we have discovered between digital machines and human beings, especially in regard to their respective language capabilites, their perceptual limitations, and their powers of imagination and conjecture.

Thus, (a′) a language framework **L** needs to be substituted for a human's semiotic abilities, for all of the reasons explained in Chapter 4. If the mode of operation of the machine is that of a physical symbol system as a special kind of computational device, then use of syntactical manipulations in lieu of linguistic understanding ought to be acknowledged. Moreover, it seems to be clear that the perceptual capabilities of machines are vastly inferior to those of human beings, not least of all due to the Principle of Minimality. When a minimal difference in perceptual experience can make a maximal difference to its interpretation, the infinitely variable perceptual response of human beings cannot be matched by the finitely limited pattern recognition accessible to digital machines. For reasons of this kind, let us assume that these machines can be provided with (b′) experiential findings by humanly processed data-inputs as required [cf. Stillings et al. (1987), Ch. 12].

Since these machines are restricted in their modes of operation to string-manipulation and to syntactical operations, they are capable of processing (c′) deductive inference and (d') inductive inference, so long as they are implemented by purely syntactical procedures. Purely syntactical procedures, of course, can be meaningful to the users of a machine without having to be meaningful to that machine itself. Whether or not machines are capable of exercising (e′) the powers of imagination and conjecture, however, raises a number of difficult problems with respect to their ability to revise and improve upon the language framework **L** that they happen to employ, where the clarification of this issue, like so many others, requires the elimination of sometimes subtle ambiguities, to which we shall give consideration here.

The utilization of a language framework **L** may provide the foundation for implementing inductive and deductive rules of inference, which can be applied to experiential findings, for example, if they are introduced into the system's data-base. These distinct faculties, of course, require objectivistic interpretations, in the sense that systems accommodated with specific combinations of (a′) through (d′) should exhibit the tendency to accept, to reject, and to hold in suspense all and only those hypotheses belonging to a single set (or set of sets) of items as its knowledge. Thus, the primary concern of epistemic systems of this kind is the incorporation of the appropriate epistemic abilities by means of which knowledge content within a domain can be acquired. To differentiate these systems from ordinary expert systems, therefore, I shall refer to them as "knowledge-processing systems". There is yet another type of system that incorporates features of systems of both of these kinds, however, which may be referred to as "expert-aided knowledge-processing systems". Their respective merits warrant consideration.

The Expert System. As perhaps ought to be emphasized, expert systems and knowledge-processing systems implement very different conceptions of the processing of knowledge. According to the expert-systems concept, knowledge-based systems reflect a triadic relationship between an expert, a program, and a user. Here, the knowledge engineer mediates between the expert and the software engineer to insure that the program reflects the content that the expert has acquired from the world by means of his abilities and training. This triadic relation can be diagrammed, where the boxed portion separates the ingredients of the system itself from its users:

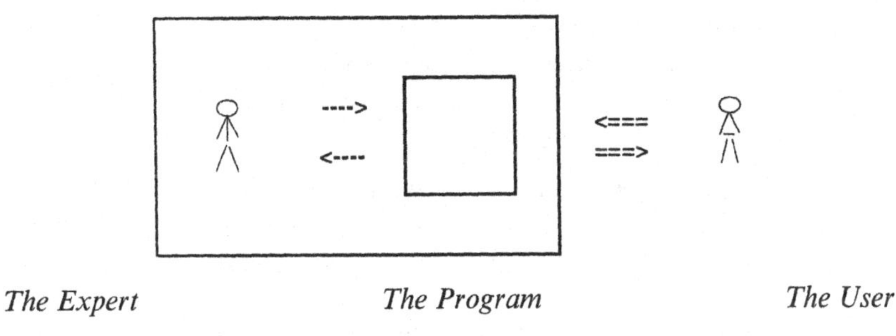

Fig. 21. The Expert System.

The two broken arrows ('--→') drawn between the expert and the program reflect the work of the knowledge engineer and the software engineer in going back and forth between the expert and the program to insure that it has the proper content, and the two broken double-arrows ('⇒') between the user and the program reflect the question-and-answer interactive exchange that enables the user to utilize that program in order to solve his problems.

The success of an expert system of this kind essentially depends upon the extent to which the expert has been able to acquire knowledge by his own education, training, and interaction with the world. Since the knowledge engineer is working directly with the expert rather than with the world, however, what is being programmed into the machine by the combined efforts of the software engineer and the knowledge engineer is a certain set of *beliefs*, namely: those that are possessed by the expert with respect to his domain of expertise. But the knowledge engineer and the software engineer, in their respective capacities, have no way of knowing whether or not that set of beliefs stands in the special relationship to the world that would be the case if those beliefs happen to be *true*. As a result, neither the knowledge engineer nor the software engineer have any way to tell whether or not the beliefs that they are programming properly qualify as "warranted". The two broken arrows between the expert and the program, therefore, tacitly represent a triple-play from the expert to the knowledge engineer to the software engineer, where misunderstanding at any position can be hazardous to the success of projects of this kind. The principal defects of these expert systems are those that we have already discussed, namely:

(1) the problem of selecting a suitable expert,
(2) the complete dependence upon his expertise,
(3) the need for experiential findings as input, and
(4) the potential for misinterpretations of outputs.

All four of these problems are inherent with expert systems of this kind. If any of them could perhaps be overcome, therefore, that might possibly justify resorting to expert systems of some fundamentally different kind.

The Knowledge-Processing System. If the system itself could possess the epistemic abilities of an expert with respect to its domain, however, then it might be possible to develop a novel kind of expert system that not only has the knowledge content appropriate to its domain but also the epistemic abilities. According to the knowledge-processing system concept, there is again a triadic relationship, but now it obtains between a user, a program, and the world, where the program itself has become an interactive knowledge-processing expert of its own! This relationship can be diagrammed as follows:

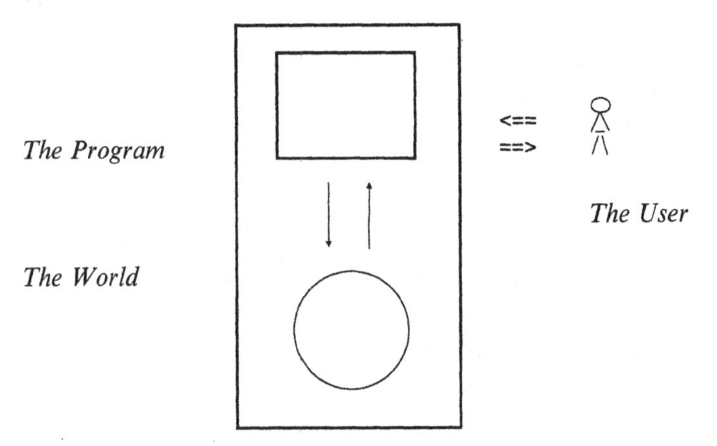

Fig. 22. The Knowledge-Processing System.

where the two solid arrows ('→') between the program and the world are intended to represent the interactive accumulation of experiential findings by means of questions and answers directed toward the world, and the two double arrows ('⇒') beween the user and the program again represent the question-and-answer interactive exchange that enables the user to utilize that program. The "expert" has been intentionally removed from the loop.

The knowledge-processing system tends to overcome at least two of the difficulties that plague ordinary expert systems, precisely because a system of this kind does not depend upon the selection of an expert for its knowledge content and is therefore not completely dependent upon his expertise. However, other problems arise in their place, especially:

(1) the adoption of some suitable methodology,
(2) the introduction of domain-specific language, and
(3) the discovery of conjectures worth testing,

where the elimination of the domain expert seems to dictate the need for a replacement expert with respect to methodology instead. Thus, while the remaining two problems appear to be virtually inescapable,

(4) the need for experiential findings as input, and
(5) the potential for misinterpretations of outputs,

the significance of the domain expert becomes increasingly apparent, insofar as his elimination creates three difficulties not easily overcome.

Indeed, the situation is even worse than has been described, for, to the extent to which a knowledge-processing system is bereft of the benefits of imagination and conjecture in introducing new predicates for the formulation of new hypotheses, it appears to be restricted in the kind of evidence that experiential findings can provide. Without imagination and conjecture, in other words, there would appear to be no way for a knowledge-processing system to rise above the plane of observation in an endeavor to ascend to the discovery of new theories that cannot be completely expressed by using observational language alone. One of the limitations of systems of this kind thus seems to be parallel to the difficulties that arise in employing the straight rule, for example, where an inference to the best explanation is required instead.

The Expert-Aided Knowledge-Processing System. These difficulties tend to disclose the multiple functions performed by experts in the development of expert systems and thereby reinforce their significance. Thus, the primary problem with expert systems is that they are overly sensitive to the problem of selection which the knowledge-processing system tends to overcome, while the primary problem with the knowledge-processing system is its inability to compensate for some of the losses that attend the elimination of the expert from the loop. Perhaps a compromise can be struck that preserves the strengths of approaches of both kinds while overcoming the respective weaknesses they display.

In order to appreciate the benefits of expert-aided knowledge processing systems over ordinary expert systems, therefore, it is crucial to realize that the knowledge content that an expert possesses has been acquired by the use of his own epistemic resources. When a knowledge engineer aims at acquiring an expert's knowledge content, in other words, the knowledge that knowledge engineer derives – when he is successful in its extraction – was first acquired by that expert by means of his own procedures. Since any expert must be an instance of *some* knowledge-processing system, the problem of selection cannot be circumvented. When we select an "expert" on the basis of sociological criteria and thereby assume that his knowledge is *expert*, we are tacitly taking for granted that his knowledge has been acquired by means of appropriate epistemic procedures. But when we select an "expert" on the basis of methodological criteria, by contrast, we then are explicitly evaluating whether that knowledge has been acquired by means of *appropriate* epistemic procedures. Either way – implicitly or explicitly – a significant commitment to epistemic procedures cannot be circumvented.

Assuming that the choice of a suitable methodology can be given an objective resolution, the point of an expert-aided knowledge-processing system is to exploit the domain expert as a *source of ideas* while relying on the knowledge-processing system to *choose between them*. This specific division of labor, moreover, closely corresponds to a distinction within the theory of knowledge between "the context of discovery" and "the context of justification". For the source of ideas does not matter to the discovery of knowledge, so long as those ideas that are selected for adoption can be justified appropriately on the basis of inductive, deductive, or perceputal inference, as required. The original source of those ideas does not matter!

The role of the expert within this scenario, therefore, is to propose and that of the knowledge-processing system is to dispose, where the potential benefits of their combination are very great. For *the use of an expert* tends to overcome two of the main problems with knowledge-processing systems:

(a) the discovery of conjectures worth testing, and
(b) the introduction of domain-specific language.

The use of an expert *as a source of ideas* rather than of specific knowledge thus tends to overcome two of the principal problems with expert sytems:

(c) the problem of selecting a suitable expert, and
(d) the complete dependence upon his expertise.

An expert-aided knowledge-processing system assumes the following form:

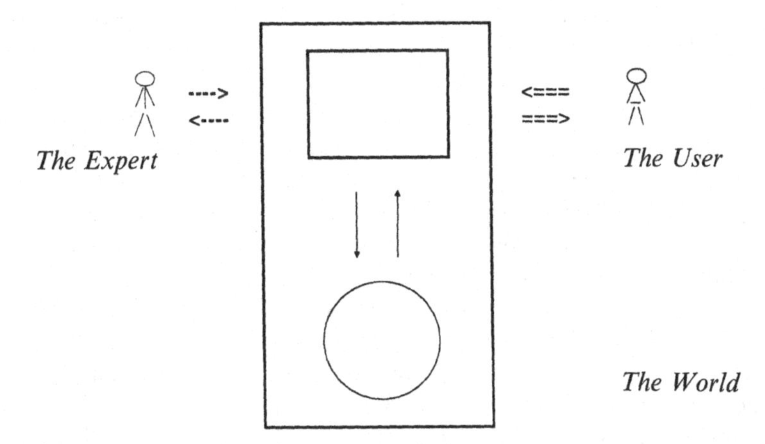

Fig. 23. The Expert-Aided Knowledge-Processing System.

where the two broken arrows ('⇢') between the expert and the program indicate the use of the expert as a source of ideas that can be processed into knowledge, while the two solid arrows ('→') indicate interaction between the program and the world in testing ideas supplied by the expert against experiential findings and the two broken double-arrows ('==⇒') indicate the question-and-answer interactive exchange that enables the user to utilize the knowledge that has been acquired by employing such a program itself.

Neither knowledge-processing systems nor expert-aided knowledge processing systems, however, are able to overcome two other problems, namely:

(x) the need for experiential findings as input, and
(y) the potential for misinterpretations of outputs.

But these are problems that expert systems themselves are not able to overcome. What hangs in the balance, therefore, is the advantage of exchanging the choice of an expert for the choice of a methodology, which is dictated by

(z) the adoption of some suitable methodology.

For it looks as though the design of a knowledge-processing system, whether expert-aided or not, confronts the same problem confronted by the design of an expert system, albeit in a different form. For the problem of the selection of an expert for his *knowledge content* now appears to have been displaced by the problem of the selection of an expert for his *epistemic skills* instead.

We have already discovered that sociological criteria of selection will be warranted, in general, if and only if the "experts" that are selected happen to employ appropriate standards of reasoning. A commitment to methodology, therefore, cannot be avoided. Perhaps the most important benefits that accrue from the knowledge-processing point of view thus derive from rendering these methodological commitments *explicit* rather than leaving them *implicit*. For when explicitly approached, the differences between alternative methodologies can be objectively assessed. And when they are objectively assessed, it becomes possible to employ them not only for the purpose of selecting "experts" for expert systems but also to build knowledge-processing systems that implement them. In this fashion, it becomes possible to create "expert systems" that can perform even better than the "experts" as systems that can acquire knowledge by justified procedures.

Since the methods that experts subjectively employ are the ones that they ought to employ only when they are objectively justifiable, all who are engaged in the design and development of expert systems should be thoroughly immersed in principles of methodology. And the domains of inquiry that pursue these principles are not difficult to discern, especially since they correspond to different kinds of knowledge. In the case of ordinary knowledge, of course, they are studied by the theory of knowledge; in the case of scientific knowledge, they are studied by the philosophy of science. As we have already discovered, both of these domains have to be pursued as *normative* rather than as *descriptive* disciplines, since the fact that certain methods are relied upon does not establish that they should be.

The principal benefit of the knowledge-processing approach, therefore, appears to be the objective conception of knowledge that it represents, insofar as "knowledge" itself can be defined in relation to a system of such a kind as *objectively warranted belief*. By comparison, the expert systems approach, at best, supports a subjective conception of knowledge in which "knowledge" itself can be defined in relation to a system of such a kind as *subjectively warranted belief*. For any expert operates on the basis of his epistemic resources, which reflect his subjectively internalized "habits of mind", let us say, which may or may not represent appropriate objective standards of reasoning. Yet it is only when an expert's subjectively internalized habits of thought correspond to objectively justifiable standards of reasoning that the expert's knowledge content itself is suitably warranted.

Concern for the objective justification of justifiable standards of reasoning, therefore, appears to be of fundamental significance for the expert-system builder as well as for the expert. Indeed, if an "expert" qualifies as an expert when his subjective habits of thought correspond to justifiable epistemic procedures, then an "expert system" qualifies as a knowledge-processing system when its own rules of inference (or "inference engine") likewise reflect justifiable epistemic procedures. And if the beliefs that are accepted by an "expert" qualify as expertise when they are the product of an acceptable epistemic process, an "expert system" will qualify as knowledge-based when the data-base for its domain has been acquired in accordance with objectively justifiable procedures. Their relations thus seem to be as follows:

	The Expert:	*The Expert System:*
Epistemic Resources:	Habits of Thought	Inference Engine
Knowledge Contents:	Accepted Beliefs	Domain Data-Base

Fig. 24. Experts and Expert Systems.

Both "experts" and "expert systems" can fail to qualify either as "knowledge based" or as "knowledge-processing systems" when they are not justifiable.

The question that naturally arises at this point concerns the existence of expert-aided knowledge-processing systems in this sense. It would therefore be appropriate to conclude this discussion of expert knowledge-based systems by pointing to an example that provides a fascinating illustration of the implementation of a knowledge-processing system of this kind. The work that best exemplifies this approach has been done by Donald Michie and Ivan Bratko in developing programs for automating knowledge acquisition along these lines [Michie (1985) and (1986); Michie and Bratko

(1986)]. Having discovered that experts tend to be very good at identifying samples and examples within their respective domains but rather poor at articulating their reasoning, Michie uses an automated process of "rule induction" as a method for discovering the generalizations that are implicit in their work.

Michie's motivation was practical rather than theoretical, since his objective was to overcome the problem generated by the difficulty that experts encounter in characterizing their knowledge for the benefit of knowledge engineers [Michie (1986), p. 9]. First identified by Edward Feigenbaum, it has come to be known as the "Feigenbaum bottleneck". There are excellent reasons for appreciating its importance, however, since, as we discovered in Chapter 3, the primitives by means of which the linguistic ingredients of declarative knowledge are ultimately grounded derive their meaning as a function of the habits, skills, and dispositions that create a bridge between language users and the world. If knowing how is the foundation for knowing that, there is every reason to suppose it must be difficult to put into words.

The system devised by Michie builds upon the ID3 algorithm fashioned by Earl Hunt and Ross Quinlan, which they employed to discover end games in chess [cf. Quinlan (1983)]. Given a collection of positive and negative instances, where each instance is the description of an object in terms of a fixed set of *reference properties*, the ID3 algorithm produces a decision tree for differentiating positive from negative instances [Michie (1986), p. 12]. The expert's role within this scheme of things thus becomes that of specifying a fixed set of reference properties and of identifying positive and negative instances of the *attribute of interest*. The expert thereby isolates a spectrum of causally relevant properties, which, if complete, permits the algorithm to discover maximally specific inductive generalizations about that attribute. The generalizations that result from this process thus have the features of lawlike sentences when those tests are severe [cf. Fetzer (1981), Chapter 9].

There are several keys to the successful implementation of this program for the purpose that it is intended to serve. One of these is that relying on experts to guide the process by identifying positive and negative instances produces a conjectural (or "prototypical") scientific theory. Specification of some finite set of causally relevant properties whose presence or absence might make a difference to the occurrence of those attributes, after all, is one form of theorizing. Another of these keys is the use of rule induction to process these samples against a background of relevance relations in the pursue of generalizations. Given predicates that designate those properties whose presence or absence makes a difference to the truth of these generalizations, the processes that are employed appear to be justifiable from the perspective of the procedures appropriate to the search for laws in science.

While Michie and Bratko adopt Hunt and Quinlan's ID3 algorithm specifically as the foundation of their own rule-induction method, an objective justification of these procedures as principles of inductive reasoning ultimately depends upon the demonstration that their use is an appropriate means for achieving the purpose for which they are intended. In the case of deductive rules of inference, the problem is to demonstrate that they are truth-preserving. In the case of inductive rules of inference, the problem is to establish that they are knowledge-expanding. The discovery of the existence of this work on the acquisition of knowledge by Michie and Bratko has been especially gratifying, therefore, not merely because it exemplifies the potential for expert-aided knowledge processing systems, but also because the method of rule induction upon which it is based bears a striking resemblance in spirit, if not in letter, to the *Inference Rule for Inductive Acceptance* that has been an element of my characterization of induction [Fetzer (1981), Part III].

It therefore appears as though the theoretical justification for the implementation of specific rules of reasoning as the epistemic foundation for an expert system of this kind ultimately depends upon and varies with the extent to which those rules of reasoning themselves can be justified. If there is a case to be made for the fundamental role of the theory of knowledge in relation to the theory of knowledge representation – as I have argued in the last several chapters of this book – then it could not be more evident than it is in cases of this kind. For the success of rule induction by systems of this sort clearly depends upon and varies with the inductive rules of inference which they employ. But the justification of those rules of inference themselves falls within the pale of the theory of knowledge. The conclusion that the proper foundations of the theory of knowledge representation are to be found in the theory of knowledge thus appears to be very difficult to resist.

PART III

REPRESENTATION AND VERIFICATION

7. KNOWLEDGE REPRESENTATION

Ultimately, the key to understanding the character of knowledge representation requires the recognition that knowledge-representation schemes – such as semantic networks, scripts and frames, and the like – have two sides or aspects, which have to be coordinated or synchronized for representation to succeed. Adequately understood, knowledge representation techniques must involve both specific logical structures and specific epistemic content:

(I) Knowledge representation requires:
  (A) logical structures reflecting:
    (1) choice of vocabulary,
    (2) formation rules, and
    (3) transformation rules; and
  (B) epistemic content reflecting:
    (1) choice of domain,
    (2) level of analysis, and
    (3) sources of information.

Establishing a suitable foundation for comparing different methods of representing knowledge, therefore, entails discriminating between these different dimensions of knowledge representation. These comparisons may vary depending upon whether (A) differences in logical structure or (B) differences in epistemic content are utilized as the relevant basis for their comparison.

As we discovered in Chapter 4, the beliefs of a person z at a time t tend to qualify as *knowledge* as a function of the extent to which they satisfy at least four conditions. Two of these – namely, the consistency and the closure conditions – are requirements whose satisfaction can be ascertained by purely deductive procedures. Given the beliefs accepted by z at t, the rules of deductive reasoning are sufficient to determine whether or not these requirements have been satisfied. But two of these – namely, the partial evidence and the complete evidence conditions – are requirements whose satisfaction cannot be ascertained by purely deductive procedures, since the extent to which these conditions are satisfied depends upon the application of inductive rules in addition to the application of deductive rules of reasoning.

When these rules are applied to a set of beliefs *B* represented by a set of sentences *S* within a language framework **L** for *z* at *t*, it becomes possible to appraise the extent to which those beliefs satisfy the conditions appropriate for *rationality of belief*. Even then at least two difficulties – which are important and pervasive – yet remain. Given an absence of direct access to the truth, there is no way to differentiate rational beliefs that are true from rational beliefs that are false. And this extends to the results of perceptual inference in the form of experiential findings: in the absence of direct access to any perceptual truths, there is no way to differentiate perceptual beliefs that are true from perceptual beliefs that are false. Even the satisfaction of these four conditions for rational belief does not guarantee the truth of the beliefs accepted by *z* at *t*, but instead supports their conditional rationality.

The notion of *conditional rationality* may be utilized in order to take into account that perceptual inference arises from a causal relation between a person *z*'s semiotic abilities at *t* and his capacities and opportunities to exercise those abilities at *t*. When a person hears certain sounds, tastes certain flavors, smells certain aromas, sees certain sights, or feels certain textures, an act of cognition takes place involving a causal interaction of his abilities, capabilities, and opportunities. The beliefs that result from these cognitive occurrences might or might not be represented by mental tokens that have the form of sentences in a natural language, such as English. But whether they are represented by icons, indices, or symbols, there are no built-in guarantees that the beliefs that have thereby been acquired must be true.

Indeed, we are aware of many circumstances under which the results of an act of cognition could easily result in false beliefs. Persons who are hard of hearing, who have poor vision, etc., are instances of systems that do not function as standard perceptual systems relative to the conventional norm. Establishing *the conventional norm* can be accomplished by the selection of specific outcomes under specific conditions as those defining the norm (hearing tests and reading tests under standardized conditions are relevant illustrations). Persons who are not hard of hearing, who do not have poor vision, etc., may function as standard perceptual systems relative to the conventional norm, but they do not necessarily function successfully. Thus, they might have been blindfolded, have cotton in their ears, etc., because of which they might not undergo the acts of cognition that otherwise would have occurred.

Moreover, even when persons do not suffer from impairments to their faculties of sense-experience because they lack the necessary abilities and are not otherwise inhibited from exercising them because they lack corresponding capabilities, this does not guarantee that the opportunities that are occasioned by the presence of things that can stand for things must lead to accurate perceptions. For human beings, as semiotic systems, are capable of making mistakes. Even when person *z* at time *t* can detect and identify things of kinds *K*1, *K*2, ..., and specific instances of things of kinds *K*1, *K*2, ... are within suitable causal proximity of person *z* at time *t*, that does not guarantee that person *z* at time *t* must therefore infer their presence. The existence of a gap between *having an experience* and *interpreting that experience* – "the problem of perception" – is one that can never be overcome.

When the rationality of belief for a person *z* at a time *t* is subject to an assessment in relation to conditions such as (CR-1) through (CR-4), therefore, the result of that assessment has to be assumed to be an assessment of conditional rationality. For the truth of the experiential findings that *z* accepts at *t* ordinarily has to be taken for granted. Thus, if we can assume that the singular sentences that describe ordinary objects that occur within *z*'s set beliefs are true, then we can assess the extent to which the other beliefs that *z* accepts are suitably supported – deductively or inductively – relative to those experiential findings. When the truth of these experiential findings cannot be taken for granted, the rationality of *z*'s entire set of beliefs becomes a problem that cannot be resolved by logic alone. But by exploring the origin of various beliefs, its existence can still be discovered.

One means that is available for resolving questions of this kind (whether with respect to perceptual findings or with respect to other beliefs that are accepted on their foundation), therefore, appears to be that of observation and experimentation intended to replicate the experiences leading to them. When *z* believes that there was poison in her tea at *t*, the best way to determine whether that perception was accurate or not would be to locate what *z* took to be her tea and ascertain whether or not it was poisoned. The difficulties in doing so are too obvious to belabor, including, for example, the possible disposal of the sample between its being served at *t* and its being tested at *t**, the poison being added after its being served at *t* but before it is tested at *t**, and all the rest. The circumstances that would make it possible to replicate the conditions that led to a perception can be problematical. But this does not diminish the importance of replication to empirical inquiry.

What all of this means is that the problem of perception reflects the uncertainty of synthetic knowledge: even the results of observations and experiments have to be entertained as findings that are fallible! Apart from the special classes of logical truths and of analytic sentences, whose truth can be ascetained *a priori* in relation to a language framework **L**, we do not possess any "knowledge" that could not possibly be false. When we employ that term to characterize any set of beliefs *B* as "knowledge", therefore, we run the risk that those beliefs might possibly be untrue. The rationality of belief for *z* at *t*, therefore, ordinarily reflects a conditional judgment that is predicated upon the assumption that the experiential findings that *z* accepts at *t* are true. For generally reliable perceptual processing appears to be a necessary condition for rendering significant assessments of the extent to which the partial and the complete evidence conditions have been satisfied.

These problems in the theory of knowledge, however, are problems for the theory of knowledge representation only when what is intended to be represented is supposed to be knowledge in the sense of *warranted beliefs that might possibly be true*. When the objective is merely to represent the set of beliefs *B* accepted by a person *z* at a time *t* (where that person might be of special interest, for example, as an expert, a politician, or a patient) or of a community *Z* at a time *t* (where that specific community might be composed of scientists, students, or voters, for example), the extent to which the requirements for rational belief (CR-1) through (CR-4) have been satisfied is a matter of secondary importance. When the beliefs to be represented are those of a madman, there is no reason to think that they should be rational.

SEMANTIC NETWORKS

Thus, there are many excellent reasons to infer that the AI community is less concerned with the represention of "knowledge" than with the representation of "belief". As William J. Rapaport, for example, has observed,

> The distinction between knowledge, in particular, and beliefs or thoughts, in general, is an important one, for one can think about things that do not exist and one can believe propositions that are, in fact, false.... But one cannot *know* a false proposition. Yet, if an AI system is to simulate (or perhaps be) a mind or merely interact with humans, it must be provided with ways of representing nonexistents

and falsehoods. Because belief is a part of knowledge and has a wider scope than knowledge, the term *belief representation* is a more appropriate one in an AI context. [Rapaport (1986), p. 372]

Pursuing this objective within the context of AI, Rapaport has collaborated with Stuart C. Shapiro in building on the work of Alexius Meinong and of Hector-Neri Castaneda [Rapaport (1985), (1986); Castaneda (1967), (1970), and (1972)]. The result of this effort is an application of the semantic network known as "SNePS" [Shapiro (1979) and Shapiro and Rapaport (1987)].

The purpose of this section, however, is not to investigate the intriguing aspects of this specific program but to explore the features that tend to distinguish semantic networks in general, especially from the point of view of the logical structure that these schemes implicitly reflect. SNePS, for example, is propositional, while other networks are taxonomic. The most striking feature of semantic networks from this perspective is their flexibility in the representation of sets of beliefs – even when those beliefs fail to satisfy the consistency and closure conditions, much less those of partial and complete evidence. When the conditions that are imposed by (CR-1) and (CR-2) are viewed as deductive requirements (of "minimal rationality", let us say) and the conditions that are imposed by (CR-3) and (CR-4) are viewed as evidential requirements (of "maximal rationality"), then semantic networks can be used to represent beliefs even when they are not even minimally rational.

A typical example of a taxonomic network would be a related system of nodes and links (or "arcs"), where *nodes* designate specific objects or properties those objects might possess and *links* of various kinds represent various sorts of relations that may obtain between those objects or properties. When they are diagrammed, spatial relations are frequently employed as a device for displaying orders of relations between things of these kinds. If a node stands for a specific object *c*, for example, a higher node may stand for a property of that object and yet higher nodes might stand for properties of those properties, etc., creating a hierarchy of relations between properties. [Some very useful discussions of semantic networks include Quillian (1967); Woods (1975); Brachman (1979) and (1983); and Maida and Shapiro (1982).]

A very simple example might involve the use of various predicate designators for *bird*, for types of birds, such as *canary* and *ostrich*, and names for individual instances of any of those types, such as *Tweety* and *Marty*. Then a semantic network could be cast that characterizes their relations as follows:

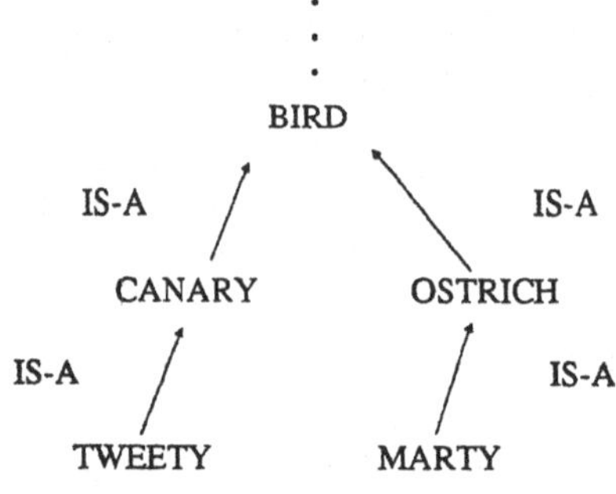

Fig. 25. A Semantic Network.

Thus, these single arrows reflect property-instantiation relations when read from bottom up. The intended interpretation of Figure 25, therefore, is that "Tweety" and "Marty" are the names of two individual things; that "canary", "ostrich", and "bird" are predicates designating properties, which individual things can instantiate; that Tweety is a canary; that Marty is an ostrich; and that Tweety and Marty are birds, which, presumably, could be explained by reference to Figure 25, because canaries and ostriches are (species of) birds.The upward series of dots, of course, indicates that this could be expanded.

There are several respects in which semantic networks of this kind can be improved. In the first place, the presumption has been made that this network represents properties and their instances, but they are frequently employed to represent classes and their members instead (where properties are intensional, while classes are extensional, kinds of entities). In the second place, the IS-A relation is being employed here to represent (what might better be viewed as) two different kinds of relations, which makes it ambiguous (or "overloaded"). The lower IS-A arrows, for example, stand for class membership (or property instantiation) relations, while the upper IS-A arrows stand for class inclusion (or property manifestation) relations. Whether nor not these distinctions are drawn depends upon their purpose; individual semantic networks can be designed to fit a user's specifications.

A different (stratified) version of the same semantic network, for example, may potentially overcome at least some of these difficulties by making explicit those hierarchical relations that are implicit in Figure 25, as follows:

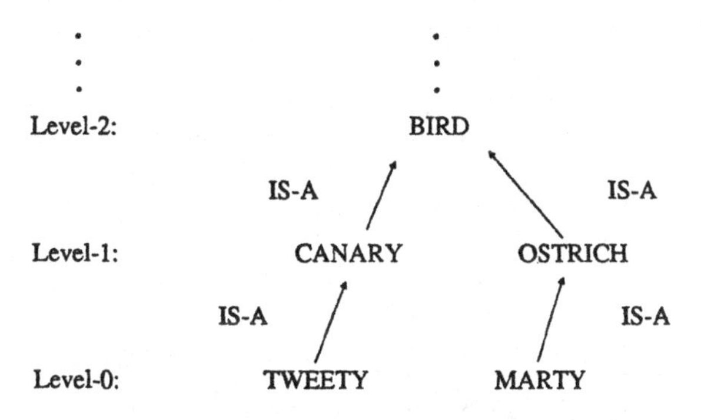

Fig. 26. A Stratified Semantic Network.

where these single arrows reflect property instantiation relations and Level 0 reflects the level of individual things, Level-1 of properties of things that are of Level-0, Level-2 of properties of things that are of Level-1, etc., in the spirit of Aristotle's theory of predication, which was developed on the basis of biological relations [cf. Jaeger (1960) and Ackermann (1965), Part 2], or of Bertrand Russell's "theory of types" [cf. Russell (1919) and Church (1959a)].

Semantic networks, moreover are highly versatile. As Hector Levesque (1986a), for example, has noted, relations between married couples may be represented by semantic networks as easily as those between other things:

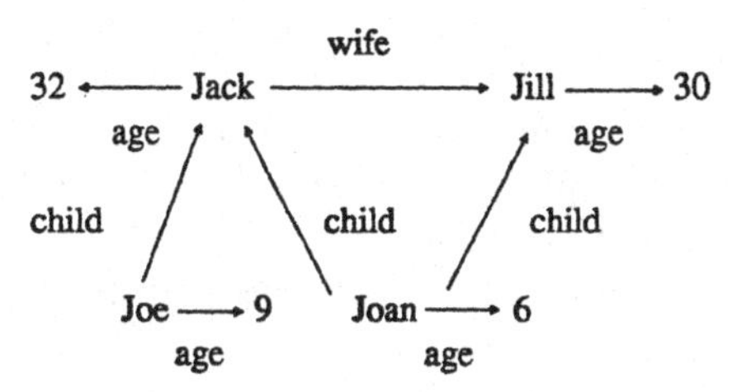

Fig. 27. A Family Semantic Network I.

Since the symmetry of the marriage relation implies that if x is married to y, then y is married to x, an equally informative diagram might seem to be:

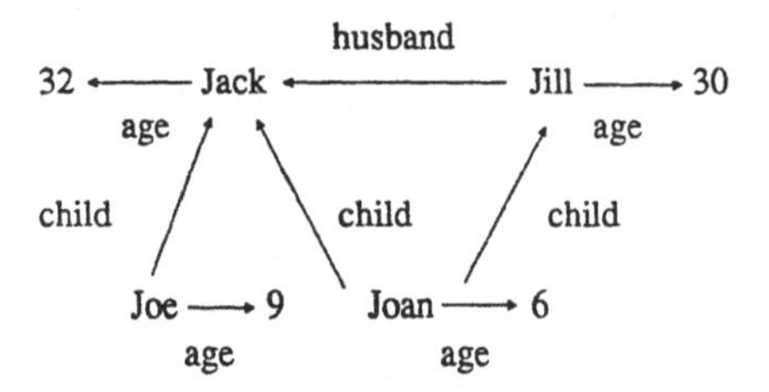

Fig. 28. A Family Semantic Network II.

Both of these semantic networks seem to convey basically the same information, namely: that Jack is the husband of Jill and that Jill is the wife of Jack, that Joe and Joan are children of Jack, that Joan is a child of Jill, etc.

Nevertheless, it is important not to read too much out of networks of this kind by importing background knowledge or beliefs as a function of the user's own context. It would be easy to infer, for example, that Jill is 30 *years* old; that Jack has a *son*, Joe, who is 9 *years* old; that Jack and Jill have a *daughter*, Joan, who is 6 *years* old; etc. Indeed, even the *married to* relation is not displayed by these specific networks. Because husbands have wives and wives have husbands as those terms are customarily used in ordinary English, anyone aware of the standard meaning of these terms would be sorely tempted to arrive at such conclusions. But the meaning of terms within such networks cannot be taken for granted. They need to be fixed by an antecedently defined syntax and dictionary and an interpreter that implements them properly [cf. Woods (1985) and McDermott (1981a)].

Semantic networks can also be employed to represent more subtle relations between, for example, persons and their beliefs, which tend to exemplify properties that are somewhat surprising. In particular, certain types of uses of the notions of belief and desire display the properties that have come to be regarded as distinctive of "intensional contexts". Consider, for example, the various relations that may obtain between these sentences:

(II) (1) Bob believes that Jill is married to Jack.
(2) Jack is the father of Joe.
(3) Bob believes that Jill is married to the father of Joe.

Then even when (1) and (2) are both true, it might be the case nevertheless that (3) is false. Even when Bob believes that Jill is married to Jack and it is the case that Jack is the father of Joe, that does not imply that Bob believes that Jill is married to the father of Joe, because that might have been something concealed from Bob, that Bob was never told, that Bob soon forgot, etc.

Although cases of this kind reflect the most commonly cited cases of intensional contexts, there are at least two other kinds of cases that ought to be identified. The first arises when someone possesses the relevant information to draw a related inference, but nevertheless fails to do so. Such a case would occur if Bob were to believe that Jill is married to Jack and also to believe that Jack is the father of Joe, yet fail to draw a related conclusion. Consider, for example, the following three sentences in comparison with (II):

(III) (1) Bob believes that Jill is married to Jack.
(2) Bob believes that Jack is the father of Joe.
(3) Bob believes that Jill is married to the father of Joe.

Even when (1) and (2) are both true, (3) might nevertheless be false, as a result of Bob's limited powers of reasoning, neglect to think things through, etc.

The second of these two additional kinds of cases occurs when someone possesses the relevant information to draw a related inference, yet nonetheless, perhaps on independent grounds, believes something incompatible with his other beliefs. Consider the following sentences in comparison with (III):

(IV) (1) Bob believes that Jill is married to Jack.
(2) Bob believes that Jack is the father of Joe.
(3) Bob believes that Jill is not married to the father of Joe.

In cases of this kind, even when (1) and (2) are both true, (3) might still be true as well. Presumably, either Bob puts (1) and (2) together and comes to believe that Jill is married to the father of Joe or he does not. Either way, as long as he continues to believe that Jill is not married to the father of Joe, he possesses beliefs that (implicitly or explicitly) cannot be simultaneously true.

Intensional contexts are commonplace, including as they do all cases of z believing (desiring, hoping, fearing, wishing, dreading, etc.) that a situation p is the case. But various techniques can be employed to represent contexts of these kinds by corresponding link-signs for belief (desire, hope, etc.). SNePS, for example, copes with some of these problems by trading in arc relations for propositional nodes. Instead of diagramming the propositional content of a belief (that Jack is married to Jill) by means of arcs and nodes, a node is used (representing the proposition that Jack is married to Jill) and a belief arc points to it. The SNePS network thus permits the display of a full range of intensional contexts of different kinds, including beliefs *de re*, *de dicto*, and *de se* [see Rapaport (1986) for a lucid introduction]. The potential for semantic networks to represent states of knowledge and belief for specific individuals is an important prospect, no doubt, especially when the aim is to represent the states of knowledge and belief for more than just one person.

Logical Structure. From the point of view of logical theory, the prinicipal ingredients of semantic networks are proper names and *n*-adic predicates. *n*-adic predicates, in turn, may be unpacked as *n*-adic *sentential functions*, which are phrases and expressions that turn into sentences when they are appropriately related to a corresponding number *n* of names. The scheme whereby sentential functions are related to names is a system of links and nodes. Some nodes are proper names standing for specific individuals, but others are *n*-adic predicates that stand for properties and relations. A variety of other denotative devices can also be utilized in lieu of proper names. Links are *n*-adic predicates that stand for properties and relations that can obtain between the things that are thereby linked, where these varieties of linkage may be of various kinds that correspond to that particular domain.

The semantic network represented by Figure 29, for example, is built out of nodes that, in this particular instance, are restricted to names for persons ("Jack", "Jill", "Joe", and "Joan") and names for ages ("6", "9", "30", and "32").

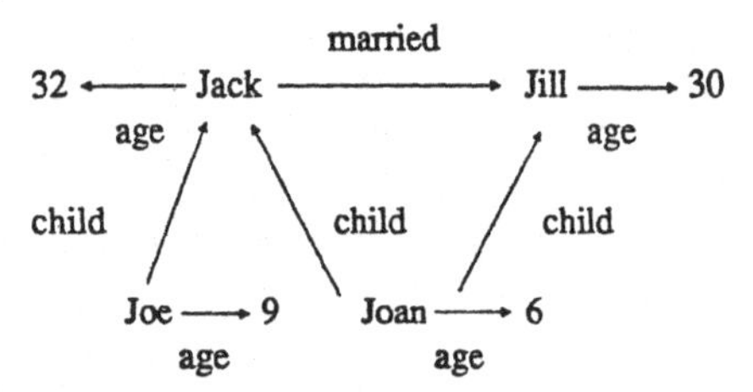

Fig. 29. A Family Semantic Network III.

Its links include four dyadic links concerning age ('___ is of age . . .'), threeconcerning biological relations ('___ is a child of . . .') and one concerning legal relations ('___ is married to . . .'). When an *n*-adic link is suitably related to a corresponding number of nodes, then a sentence will result, such as, in the case of Figure 29, "Jack is married to Jill", "Jack is of age 32", and so on.

The choice of specific denotative and designative devices depends on the domain the semantic network is intended to reflect and the level of analysis appropriate to its purposes. If the knowledge or beliefs that are to be represented are those of a person *z* or those of a community *Z*, then its vocabulary should include *names* or *descriptions* for every object of knowledge or belief and *predicates* and *relations* for every property and relation that is known or believed to hold among them. The resources that are needed can be diverse:

(V) A semantic network may include:
- (A) *denotative devices* such as:
  - (1) proper names: *a*, *b*, *c*, ...;
  - (2) definite descriptions; and
  - (3) functors: *f*1(*a*), *f*2(*a*), ...; and
- (B) *designative devices* such as:
  - (1) *n*-adic predicates: *Ax*, *Bxy*, ...;
  - (2) higher-order predicates; and
  - (3) formalized modes of modality.

Thus, functors are definite descriptions constructed out of proper names by means of functions (such as 'the father of ___', 'the eldest son of . . .', and the like), while formalized modes of modality include those for logical relations, lawful relations, and accidental relations (as we investigated in Chapter 4).

William Woods (1983) emphasizes what seem to be two of the most crucial dimensions of semantic networks, *expressive adequacy* and *notational efficacy*. A semantic network possesses expressive adequacy when it has the symbolic resources that are required to represent everything that needs to be represented with respect to its domain. And it possesses notational efficacy when those symbolic resources promote computational efficiency (for various types of inference), economy of representation, and ease of modification [Woods (1983), pp. 22-23]. And Hector Levesque (1986a) has suggested a criterion for the adequacy of those expressive resources, which entails the availability of names or descriptions to name or describe everything that is in that domain and predicates and relations to designate and describe every property and relation that might obtain between them within that domain. Levesque introduces the term "vivid" to characterize a knowledge base (or KB) of this very special kind. Thus, for a knowledge-based to be *vivid*,

(1) there must be a one-to-one correspondence between a certain class of symbols in the KB and the class of objects of interest in the world;
(2) for every simple relationship of interest in the world, there must be a type of connection among symbols in the KB such that the relationship holds among a group of objects in the world if and only if the appropriate connection exists among the corresponding symbols in the KB. [Levesque (1986a), p. 93]

These conditions, of course, define an *isomorphic relationship* between a set of symbols and a set of entities, where the links that connect the symbols reflect the relations between those entities. As Levesque has formulated this condition, however, it appears to be somewhat problematical, for his use of the term "simple" suggests the need for criteria of simplicity, which might otherwise be avoided. Indeed, the appropriate term here appears to be not "simple" but "relevant", where relevance in this sense is a pragmatic notion.

An important consideration that Levesque does not emphasize, however, is that an isomorphic relationship obtains with respect to a *level of analysis*. A model of the battleship *Missouri* might be a very good model, with parts corresponding to its turrets, its superstructure, its keel and hull, and so on, standing in relations to one another that correspond to those of their counterparts; yet that does not mean there are parts corresponding to the paint on its bow and rust on its anchor, not to mention the mobility and firepower of the real thing. Without a determination of which features within a domain are the relevant features, in other words, it is no more possible to ascertain whether a set of symbols for the representation of knowledge satisfies these requirements than it is to ascertain that a set of parts for a model of a ship has fulfilled them. And this requires a pragmatic determination relative to the aim, objective, or goal that this scheme or that model is meant to satsify.

This means that the logical resources which semantic networks must be able to provide need to include the apparatus necessary to draw every distinction that might be required to capture the relevant entities, properties, and relations that may obtain within that domain. Most importantly, these resources should encompass every mode of modality that could be required to draw relevant distinctions between the various ways in which objects and properties can be related. These should therefore make it possible to differentiate between properties that things may possess in three different ways:

(VI) (a) as matters of definition,
(b) as matters of accident, or
(c) as matters of natural law.

There need to be kinds of IS-A links and other types of arcs and nodes corresponding to each type of relation between properties that might be relevant to a specific network. Some networks allow users to define their own.

Ronald Brachman has identified a rather large class of alternative kinds of IS-A links, including nodes for individual objects and higher-order generic nodes [Brachman (1983), p.34]. So far as I have been able to discern, however, the most fundamental kinds of links that he identifies – including membership (in class) links, predications (of property) links, and conceptual containment (of meaning) links – correspond to the ones represented by (a)-(c). We have already discovered the differences between logical and lawful necessities, between deterministic and indeterministic causation, and between material conditionals and relative frequencies. Those distinctions provide a semantic foundation that justifies the introduction of corresponding syntactical notations for each of them, respectively, that can be utilized as appropriate with respect to choice of a specific domain, the level of analysis, etc.

The problems that confront the use of semantic networks with respect to formation and transformation rules, therefore, reduces to that of inventing or creating an appropriate (or "effective") notation for representing these distinctions by means of syntactically distinctive type-of-link signs. If we take *X* and *Y* to represent the nodes connected by those links, then consider:

(VII) (A) to represent modes of necessity:
(1) logical necessities: $X = \Box \Rightarrow Y$;
(2) lawful necessities: $X \Rightarrow Y$;
(B) to represent modes of causality:
(1) deterministic (*u*): $X = u \Rightarrow Y$;
(2) probabilistic (*n*): $X = n \Rightarrow Y$;
(C) to represent modes of coincidence:
(1) material conditionals: $X \rightarrow Y$;
(2) relative frequencies: $X - f \rightarrow Y$;

where the employment of one or another of these different types of links is dependent upon and varies with the epistemic contents that they represent.

Epistemic Contents. The epistemic content that a semantic network is intended to represent depends on a choice of domain, a choice of the level of analysis, and a choice of sources of information. The selection of a madman would be an interesting choice for any number of reasons, but it would be a mistake to imagine that the contents of his set of beliefs should be similar to those of ordinary human beings. And there are several ways in which, for example, a mentally bewildered person could deviate from the normal. One might be with respect to the specific contents of their beliefs in taking some things to be true that most of us take to be false. Another, of course, might be to believe in the existence of objects, properties, and relations we otherwise reject. Yet another might be to use the same words with different meanings than the rest of us assign them, not as a function of attitudes and emotions but instead of intensions and their corresponding extensions. The very same specific words, of course, could appear in a variety of different semantic network with very different meanings without the sources of the data for such representations having to be insane. Thus, for example:

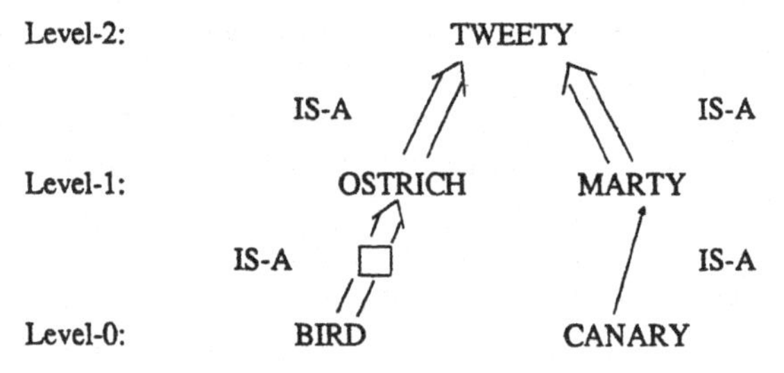

Fig. 30. An Enhanced Stratified Semantic Network.

where the various arrows reflect property-instantiation relations when read from bottom up. The intended interpretation of Figure 30, therefore, is that "Bird" and "Canary" are the names of individual things; that "ostrich", "marty" and "tweety" are predicates designating properties that individual things can instantiate; that Bird is an ostrich; etc. Then the IS-A of logical necessity that links Bird and ostrich indicates that "Bird" has been introduced as the name of a (distinctively different) ostrich; whereas the the IS-A of material conditionality that links Canary and marty indicates that "Carnary" has been introduced as the name of a (distinctively different) marty; and the IS-A's of lawful necessity indicates that those properties are linked by laws of nature.

Even in the case of semantic networks like Figures 26 and 30, therefore, it is important to realize that the interpretation of a semantic network that has been implemented by means of an interpreter and inference package is not simply a function of the context that we bring to it. In relation to our ordinary language, we tend to presume that "Tweety" and "Marty" are proper names rather than common nouns; that "canary", "ostrich", and "bird" are common nouns rather than proper names; and that "canary" and "ostrich" are common nouns that designate species of the genus designated by "bird". These background beliefs, in other words, are part of the *context* that each of us brings with us to the interpretation of any semantic network. Indeed, the extent to which our context tends to condition our interpretation is also displayed by our disposition to capitalize certain words but not others, etc.

As Woods (1975) and McDermott (1981a) emphasize, however, semantic networks specifically, like programming languages generally, come with a built-in "context" of their own when they are properly implemented. Constraints are imposed upon their possible interpretations *by the extent to which the relations between their arcs and nodes are fixed by interpreters and inference packages.* They thus exemplify completely syntactical inferential networks of the kind we explored in Chapter 3. It should therefore not be assumed that a specific network implements particular relations between its arcs and nodes merely because the names and predicates which occur within them happen to be familiar. Unless a user possesses reliable information concerning their specific design and mode of implementation, they ought to be approached as though they represented a foreign tongue.

In the case of Figure 30, for example, we should not be confident that the use of ordinary language would provide a reliable guide. For words such as "BIRD" and "CANARY" are not ordinarily used as proper names but rather as common nouns. Words like "MARTY" and "TWEETY" are ordinarily used as proper names rather than as common nouns. Yet while "OSTRICH" is normally used as a common noun, the relations that seem to be reflected between things of these kinds are not completely inappropriate. It therefore looks as though *either* these words are not being used in their ordinary senses at all, but as elements of a different language framework, *or else* these words are being used in their ordinary senses, but as elements of a peculiar – perhaps completely abnormal – set of beliefs. Which is which depends upon exactly what domain and/or source is supposed to be represented by this network.

There are, of course, some general guidelines that can be helpful in cases of this kind. The *Principle of Charity*, as it is known, suggests that, in order to render a suitable interpretation of the meaning of the words that are employed by others, an effort must be made to maximize the truth of whatever claims are made or whatever sentences are used. This principle appears to support the view that the beliefs being represented are those of someone using a different language framework. The *Principle of Humanity* suggests further that, in order to render a suitable interpretation, the assumption should be made that the motives, beliefs, and other factors that govern the behavior of others are as much like our own as they could possibly be (without violating the conditions for rationality of belief). This too suggests that the set of beliefs being represented has likely been expressed in an unusual language.

This interpretation, of course, is aided immeasurably by the stratification of this semantic network, which serves to fix some of the most important sematically relevant information about these words, especially in combination with our enhanced syntax for distinguishing kinds of IS-A links. This point, moreover, reflects a perfectly general circumstance that clearly deserves to be emphasized within the context of this inquiry. For, as we are discovering here, as we have discovered before, and as we will discover again, the distinctions that we tend to draw in our syntax are just those that contribute toward the attainment of our semantically and pragmatically driven goals. It should therefore come as no surprise when the systems of syntax we design tend to fit the distinctions that matter with respect to the meanings we would use them to convey and the purposes we would use them to achieve.

It should be observed that even additional syntactical distinctions are required to improve the potential of semantic networks as a device for representing knowledge. For the sets of beliefs reflected by Figure 30, for example, concerning Jack and Jill and their children, Joe and Joan, represent transient rather than permanent property relations that are vulnerable to change. An aspect of this phenomenon, of course, is that sentences concerning someone's age are obviously time-dependent. Another indication of its breadth and its depth, however, is that marriage can be a sometime thing: divorce, as we all know, is as American as apple pie. A further enhancement that would tend to alleviate these problems, therefore, would be to introduce temporal variables and temporal constants whenever they are necessary to convert what might otherwise remain occasion sentences into their eternal sentence forms:

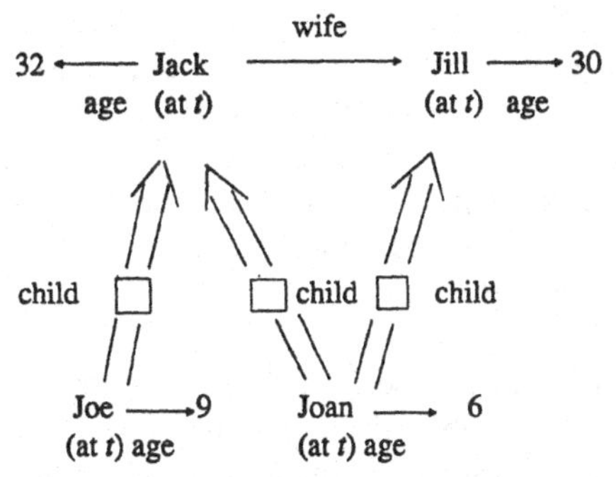

Fig. 31. An Enhanced Family Semantic Network.

where this semantic network can be enhanced as illustrated but cannot be stratified, since all of these relations obtain between individuals of Level-0. [On representing time, see esp. Allen (1983) and Almeida (1987); on belief revision, see Doyle (1979), Martins (1987), and Martins and Shapiro (1988).]

SCRIPTS AND FRAMES

One of the appealing features shared by semantic networks and scripts and frames is that they afford a relatively flexible framework for representing beliefs. First introduced by Roger Schank and Robert Abelson (1977), scripts have been very influential and widely discussed. Schank has given a description of scripts as follows: "Scripts organize all the information we have in memory about how a commonplace occurrence (such as going to a restaurant) usually takes place. In addition, scripts point out what behavior is appropriate for a particular situation" [Schank (1982a), p. 101]. Notice several features of Schank's conception: (1) they provide a means for representing the organization of information from a person's memory; (2) they apply to kinds of situations, types of occurrences, etc., that are ordinary or typical; and (3) they (perhaps tacitly) convey instructions regarding proper or normal conduct within the context of the situations to which they apply.

An analogous characterization has been offered by Charniak and McDermott, who emphasize the function of "script variables" in their specification:

> A script is a stereotyped sequence of events, such as the events in getting your hair cut or going to a restaurant. To oversimplify, we can represent a script as a list of event descriptions, which contain script variables. For instance, getting your hair cut (at a beauty parlor) has the script variables Beauty-Parlor, Receptionist, Beautician, Chair, Sink, ...[Charniak and Mc Dermott (1985), p. 393. For discussion, see Schank and Riesbeck (1979).]

The use of the term "stereotyped" within this context, I believe, is intended to indicate that scripts ordinarily concern sequences of events that are relatively familiar. From the point of view of a descriptive approach, this might be justified on the grounds that, in order for a script to be represented in a person's memory, he needs to have experienced that sequence of events, perhaps more than once. This claim appears to be doubtful for several reasons, however: someone could come to know a script by heart, even though he had never experienced it, because he had read it, had read about it, or the like; or even though she had such an experience only once, because she has a very retentive memory, it was an unusually pleasant experience, and so on.

A complete characterization of a haircut script, of a restaurant script, or whatever, of course, would require components of various different kinds:

Entry Conditions: conditions that, in general, must be satisfied before a specific script sequence can commence;
Results: conditions that, in general, will obtain by the time of that script's completion;
Props: types of objects whose presence is required in order for a script to be fulfilled;
Roles: types of persons who perform actions of certain kinds described by the script;
Tracks: variations in the sequence of unfolding events that a script may permit; and
Scenes: the actual sequence of events as they occur in accordance with the script.

The script thus carries the players from its entry conditions to its results, using the roles and props as elements of its scenes [cf. Rich (1983), p. 234].

Perhaps the most popular example of one of Schank's scripts is the restaurant script, in which someone, *S*, enters a restaurant, orders a meal, consumes that meal, and exits the restaurant. The "entry conditions" for this script to apply include that *S* is hungry and that *S* has money, which, in general, are supposed to be satisfied for it to apply; various "props" and "roles" including tables and waitresses, which, in general, are necessary ingredients for the performance of the script; and "results", such as that *S* is less hungry and that *S* has less money, which are, in general, its culmination. Thus, the events described by a script may be said to form a "causal chain", where this chain has a first link (the "entry conditions") and a last link (the "results"), while what transpires in between is connected to what went before (earlier events that make them possible) and to what goes on after (later events that they make possible), within the context of that script [cf. Rich (1983), p. 236].

Schank tends to spell out the details of scripts by employing some version of *conceptual dependency relations*, which provide a set of categories for the classification of actions by their types. Examples of these are the following:

ATRANS	Transfer of an abstract relationship (e.g., give);
PTRANS	Transfer of the physical location of an object (e.g., go);
PROPEL	Application of physical force to an object (e.g., force);
MOVE	Movement of a body part by its owner (e.g., kick);
GRASP	Grasping of an object by an actor (e.g., throw);
INGEST	Ingesting of an object by an animal (e.g., eat).

Indeed, Schank has offered more than one list of this kind. Its purpose is intended to be that of establishing semantically relevant categories for the classification of actions by their kind. The complexities of human behavior, however, are such that, with possible exceptions, to do justice to the meaning of different kinds of action within different contexts might well require a different conceptual dependency relation for each distinct kind of action. Some of the effects of these endeavors, moreover, border on the ridiculous, as when "Bill shot Bob" becomes *Bill using gun PROPEL bullet into Bob with value detrimental to health of -10* [Schank (1973) and esp. Schank (1975)].

While it is easy to be bewitched into the presumption that there are no scripts without conceptual dependency relations, that is not the case. As Schank and Jaime Carbonell have observed, all such units are conjectural:

> The initial choice of [primitives] to represent knowledge in a new domain is necessarily *ad hoc*. We make an initial, tentative commitment to a new set ... in the new knowledge domain. In the process of codifying the new knowledge, using the new knowledge in computing programs that process text and answer questions, and in light of new theoretical considerations, we modify, change or even replace our original choice.... We believe that this method rapidly converges upon a ... set of basic units that organizes the knowledge of the domain in a useful and enlightening way. [Schank and Carbonell (1979), p. 360]

While some such set of primitive event-types may be important in order to implement a script by means of a suitable computer program, they are not, strictly speaking, elements of scripts themselves and should not be thought to be required by them. They can be implemented in many different ways. There is no reason to think that the categories that Schank is disposed to use are superior to Aristotle's categories or even to those of ordinary language.

Whether or not scripts have to be "stereotyped" sequences of events, it is essential to a script that it contain a sequence of related events. A restaurant script in which the dining room alternately changed to a tropical jungle, a deserted island, and so on, would (at least) be no ordinary script, although it would not necessarily therefore be no script at all. An historical-narrative script, for example, might violate the course of historical events by including events that contravene the past; a science-fiction script might violate laws of nature by describing events that are precluded by the world's causal structure; and a fantasy-scenario script might even contravene the laws of logic by portraying incoherent happenings – but they might all nevertheless qualify as "scripts", so long as the players (or the period or the theme) remained constant throughout. Although scripts usually portray familiar sequences of events that happen to have numerous instantiations, they are open to adaptation for the purpose of portraying less familiar sequences of events as well.

Scripts as Sequences of Frames. Schank's suggestion that scripts can incorporate a normative as well as a descriptive dimension, moreover, deserves further consideration. In elaborating this possibility he remarks, for example,

> Knowing that you are in a Restaurant Script leads to knowing that if you ask a waitress for food, she is likely to bring it. On the other hand, we know that if you ask her for a pair of shoes, or if you ask her for food while she is returning home on a bus, she is likely to react as if you had done something odd. [Schank (1982a), p. 101]

Indeed, it is interesting to consider the relationship between conventional expectations (with respect to *normal* behavior) and ethical requirements (with respect to *moral* behavior). For the behavior that Schank has described qualifies more as violations of ordinary conventions than as behavior which is illegal, immoral, or fattening. Consider the alternative script in which, because the service is slow, you stab her with your fork when she finally brings your meal. This case would not only contravene conventional expectations but also would exhibit a violation of the moral principle that we must always treat others as ends and never merely as means, which is another matter entirely.

A principal benefit of the utilization of scripts is that they provide us with information that might be relevant to understanding situations, even though those situations may be only partially described. Consider these two scripts:

(S1) John went into a restaurant. He asked the waitress for the house special. He paid the check and left.

(S2) John went into a park. He asked the midget for a mouse. He picked up the box and left.

Schank suggests that (S1) represents a standard "restaurant script", but that (S2) represents a non-standard "mouse-buying script", where we can easily understand the former but cannot easily understand the latter [Schank (1982a), p. 102]. It appears as though this facility is subjective and varies from person to person and from time to time. Consider, for example, this script:

(S3) John drove up to the side of a building. He asked the woman for a cold pack. He took it and drove off;

where (S3) – at least superficially – resembles (S2) as another incomplete and fragmented description. Their ease of intelligibility depends upon how often we visit midgets in the park, purchase chicken from the Colonel, and so forth, as an aspect of our background life experiences. Their intelligibility depends on and varies with the *context* of beliefs and motives, etc., we bring to them. [Those who view these examples as "stories" rather than as "scripts" can revise them by substituting script variables for individual constants therein.]

Scripts in Schank's sense, moreover, bear a strong resemblance to scripts in Hollywood's sense. There are settings and scenes, players and roles, and is a sequence of events with a certain continuity. As in the moving pictures, scripts tend to be useful because certain patterns of events tend to occur on more than one occasion. Thus, various kinds of movies (romances, mysteries, and so on) depend upon the corresponding kinds of scripts (romance scripts, mystery scripts, and so forth). Since movies are made up of large numbers of distinct scenes, which, in turn, are composed of large numbers of individual frames, it is tempting to entertain scripts as sequences of (large numbers of) frames, where *frames* are specific fixed arrangements of things (players, settings, and so on) in which nothing changes. From this point of view, in other words, a script would consist of a temporal sequence of non-temporal frames, where the amount of time that passes from one fixed frame to another fixed frame indicates how long it took (how long it would take, and so forth) to proceed from the situation described by one frame to that described by another.

This suggestion does not necessarily reflect the ordinary practice in AI, where "scripts" and "frames" tend to be used interchangeably (and there is no real difference between them). But the notion of a frame is loose enough to afford the latitude that the adoption of this recommendation may require. Minsky himself, for example, characterizes frames by the following features:

> A frame is a data-structure for representing a stereotyped situation, like being in a certain kind of living room, or going to a child's birthday party. Attached to each frame are several kinds of information.... We can think of a frame as a network of nodes and relations. The top levels of a frame are fixed and represent things that are always true about the supposed situation. The lower levels have many terminals – slots that must be filled by specific instances or data. Each terminal can specify conditions that its assignment must meet. [Minsky (1975), p. 96]

It is striking how much of this depiction is satisfied by semantic networks as well. Without belaboring the point, both scripts and frames are alike insofar as both provide partial (or "incomplete") descriptions of types (or "kinds") of situations. These may or may not actually be encountered in daily life – notice Minsky's reference to "supposed situations" – but, no doubt, they are ordinarily used to portray familiar sequences of events. Most important, however, nothing about this conception of frames would rule out the prospect, if someone were to pursue it, of entertaining "scripts" as sequences of frames.

One of the fundamental difficulties within the domain of knowledge representation in general is attempting to ascertain whether one or another means for representing a person's beliefs has a greater range of applicability, more descriptive power, and such, than its alternatives. It is therefore of some interest to consider the possibility that scripts and frames might be represented by means of some other framework, such as predicate calculus. Patrick Hayes [(1973) and (1979)] has explored this possibility, where it turns out that scripts and frames can be represented in predicate calculus whether we choose to view them as the same thing (so that scripts are "dynamic frames" and frames are "static scripts") or not. Either way, frames can be characterized (as Minsky implies) in terms of "slots" and "fillers", where *slots* are variables (open features) of a situation and *fillers* are specific values (or corresponding details) thereof. Predicate calculus can be used to make this formal.

Consider, in particular, that objects (persons, things, . . .) are represented in predicate calculus by individual constants a, b, ... (for things which are specified by names) or by individual variables x, y, ...(for blanks to be filled by individual names), etc. A situation might then be said to be of kind F (a partial description), where a complete description of specific instances of that kind could be obtained by identifying what objects (persons, things, ...) a, b, ...are involved in that situation and their relations $R1$, $R2$, ...to one another. As Hayes (1979) suggests, these situations may then be characterized by means of predicate-calculus formulations by using n-adic predicates, say,

$$\text{(P1)}\ (x)[Fx \rightarrow (Ey1)(Ey2)\ldots(Eyn)(R1xy1\ \&\ \ldots\ \&\ Rnxyn)];$$

which means that, for every situation of kind F involving something x, there exists something $y1$ that stands in relation $R1$ to x, there exists something $y2$ that stands in relation $R2$ to x, ..., and there exists something yn that stands in relation Rn to x. A birthday party F for someone x, for example, typically involves a cake $y1$ that stands in the relation $R1$ of being a-birthday-cake for x, at least one other person $y2$ that stands in the relation $R2$ of being a-guest-at-the-birthday-party-for x, ..., and the like. In light of this, it becomes clear why birthday parties without cakes and guests are sad affairs.

When scripts and frames are distinguished as has been suggested above, however, predicate calculus may still be employed for their representation. In this case, of course, scripts are viewed as sequences of frames and each frame represents the situation that obtains at a different moment of time:

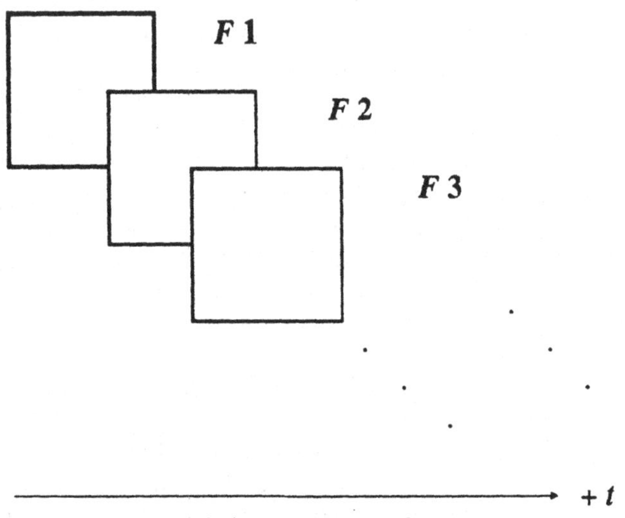

Fig. 32. Scripts as Sequences of Frames.

so that script *S* consists of the sequence of frame *F*1, followed by frame *F*2, and so on, where frame *F*1 for *x* is satisfied at time *t*1, frame *F*2 is satisfied at time *t*2 (where *t*1 is earlier than *t*2), and so forth, until the complete sequence of frames that make up that script have been fulfilled, where the notions of entry conditions, of props and roles, etc. still apply throughout. [From this perspective, Schank's "Memory Organization Packets" (or MOPs) are branching scripts of alternative sequences of frames; Schank (1982b).]

In order to capture the difference in these conceptions, therefore, the use of predicate calculus must be supplemented by temporal constants (for specific times) and by temporal variables (as blanks that are filled in by specific times). Thus, corresponding formulations may be provided for a script or sequence *S* that extends from entry conditions at time *t*1 to results at time *tn*:

(P2) $(x)(t)\{Sxt1\text{-}tn \rightarrow [(EF1xt1) \,\&\, (EF2xt2) \,\&\, \ldots \,\&\, (EFnxtn)]\}$;

which means that, for each instance of script S that transpires from time $t1$ to time tn, there exists an instance of frame $F1$ at $t1$, an instance of frame $F2$ at $t2$, ..., and an instance of frame Fn at tn. Moreover, for each such frame,

$$\text{(P3)}\ (x)(t)[F1xt \rightarrow (Ey1)(Ey2) \ldots (Eyn)(R11xy1t1 \ \& \ldots \& \ R1nxyntn)];$$
$$(x)(t)[F2xt \rightarrow (Ey1)(Ey2) \ldots (Eyn)(R21xy1t1 \ \& \ldots \& \ R2nxyntn)];$$
$$\ldots$$
$$(x)(t)[Fnxt \rightarrow (Ey1)(Ey2) \ldots (Eyn)(Rn1xy1t1 \ \& \ldots \& \ Rnnxyntn)];$$

which implies that, for each instance of script S that transpires from time $t1$ to time tn, there exists some $y1$ to yn that stands in relation $R11$ to x at time $t1$, there exists some $y1$ to yn that stands in relation $R21$ to x at time $t2$, ..., and there exists some $y1$ to yn that stands in relation Rnn to x at time tn. Results such as these support the conclusion that a temporally quantified version of predicate calculus has the potential to represent scripts and frames.

Scripts as Modes of Understanding. Whether scripts and frames are viewed as the same or scripts are viewed as sequences of frames, the difficulty arises of inferring what is true at any particular time in relation to a script or a frame. This, of course, is the problem we have previously addressed as "the frame problem", even though its existence is not dependent upon utilizing frames as a scheme for the representation of knowledge. Since "scripts" are stereotyped situations that describe "typical" or "ordinary" sequences of events, it should not be difficult to discern that they reflect the same kind of crude generalizations and "rules of thumb" that are distinctive of defeasible reasoning. Scripts are therefore vulnerable to the very same problems that affect other forms of reasoning of this kind. Inferences based upon scripts will therefore be non-demonstrative, ampliative, and non-additive as well.

One way to bring home the significance of this matter is the realization that, even with respect to a simple sequence like a restaurant script, it is important to know which is more likely to disappear first: the restaurant, its patrons, or the food? Strictly speaking, a script qualifies as reflecting a "causal chain" only to the extent to which each of the events that occur between its first and last links is brought about by preceeding links. This will be the case only if those preceeding links are instantiations of some complete set of relevant properties with respect to their successor links. Otherwise, the frames that make up those sequences will fail to be maximally specific in relation to their successor frames, in which case they do not satisfy the conditions required to bring about those following frames.

Since the satisfaction of this requirement would impose tremendous burdens upon those who would construct scripts and frames, it is important to recognize that scripts and frames, as merely partial descriptions of the relevant factors making a difference to their results, are at best only incomplete characterizations of the causal sequences to which they pertain. Indeed, in terms of familiar notions, scripts and frames must be:

(D13) open systems =df systems whose causally relevant properties are only partially specified, relative to some outcome of interest (such as a script's result).

Any systems that are not merely partially specified, by comparison, are:

(D14) closed systems =df systems whose causally relevant properties are completely specified, relative to some out come of interest (such as that script's result).

Thus, inferences that are based upon incompletely specified open systems tend to qualify as "defeasible reasoning" precisely because, as more of the slots that attend its frames are filled in, expectations that may have been initially well-founded might have to be either modified or rejected. This suggests the importance of availing ourselves of default values and of the opportunity to "change the script" as an unfolding sequence might require.

The defeasible character of the reasoning reflected by inference mechanisms implemented on the basis of scripts and frames can be illustrated by specific cases such as that of SAM (for "Script Applier Mechanism"). SAM reads stories (as inputs) in a top-down, predictive manner, drawing inferences from the information that those stories provide. Thus, for example,

> Each script provides prestored expectations about what will be read, based on whatever has already been seen.... Each of the types of texts (that SAM can read) involves certain invariant components, such as what can happen, the order in which things happen, and who is involved. This consistency of form and content enables the script-based model of reading to be used. [Cullingford (1981), pp. 75-76]

Illustrations of the types of texts that SAM can process include automobile accident reports, which it can summarize and answer questions about. Utilizing the information with which it has been previously programmed concerning the kinds of things that can happen ("frames"), the order in which they can happen ("scripts"), and the sorts of things and types of persons involved ("props" and "players"), SAM tends to complete situations that are only partially described by filling in certain of its open slots with defaults.

Strictly speaking, of course, SAM handles types of texts (or "situations") that usually involve specific invariant components, where this normal pattern allows its script-based style of reasoning to be utilized. It should come as no surprise to discover that these scripts need a branching structure with "turning points" at which the subsequent course of events must be decided:

> Turning points are places in a script where several actions might follow. In a restaurant, for example, one may take a seat or wait to be taken to a seat by an employee. Turning points also include "interference" and "resolution" activities, in which the flow of action departs slightly from the "usual". For example, all the tables may be occupied when someone arrives at the restaurant (an interference in the restaurant script), but a patron may choose to wait until a table becomes available (a resolution). [Cullingford (1981), p. 76]

Indeed, one of the lessons derived from working with SAM was the discovery of an enormous variety of "subtle, little noticed inferences" that are required to reconcile expectations based upon scripts with actual sequences of events, which may depart from them in unexpected ways. The importance of these modest but significant factors thus suggests that the Principle of Minimality applies within action contexts as well as within perceptual contexts. This is not surprising as a manifestation of the general importance of the requirement of maximal specificity. SAM's predictive powers are therefore ultimately defined by the extent to which real events conform to its expectations.

Another illustration of the defeasible character of automated reasoning using scripts is the explanation-providing program known as PAM [Wilensky (1981)]. Even though a specific sequence of events may depart from a script in real life, human beings appear to have capacities to understand that go far beyond application of simple scripts and frames. This is especially important in the case of behavior that deviates from the "normal" or the "ordinary" which occurs in stereotypical situations.

> Thus, in particular, ... story understanding involves reasoning about people's intentions. ... Reasoning about intentions is needed to find explanations for a character's actions. A reader involved in processing a text is engaged in the task of constructing an explanation for each event in that text. The reader makes inferences for the purpose of finding explanations. [Wilensky (1981), p. 136]

So PAM reads stories (as inputs) in a bottom-up, explanatory manner, like SAM drawing inferences from what those stories describe, only this time arriving at conjectures as to the motives and beliefs that brought it about.

PAM makes use of a large number of assumptions about plans and goals without which it could not undertake to explain any person's behavior. One of the fascinating discoveries that emerges from the study of SAM and PAM, moreover, is the extent to which they depend on access to expert knowledge:

> In a research effort directed at an area as large as text understanding, the only reasonable way to proceed is to design a system which, like SAM, consists of a community of experts, each accessing its own specialized knowledge base, as it tries to contribute to the problems at hand. [Cullingford (1981), p. 82]

Indeed, these reflections are especially significant with respect to our discussion of alternative frameworks for knowledge representation. For they suggest the possibility that scripts and frames, like production systems and semantic networks, can be viewed as knowledge-based systems. Whether or not these schemes differ in their logical structure and in their epistemic content, to the extent to which their success depends upon knowledge that goes beyond common sense, they are all special kinds of "expert systems".

In exploring the significance of scripts and frames, therefore, it is crucial to bear in mind their ultimate limitations. While programs based on scripts and frames such as SAM and PAM can be accurately described as inference mechanisms that are based upon "defeasible reasoning", it ought to be recognized that their capacity to predict or to explain any sequence of events to which they are applied is limited by the extent to which those events actually satisfy the assumptions upon which they have been based. When Rich extolls their "ability to predict events that have not explicitly been observed" [Rich (1983), p.236], do not overlook that what comes out had to be put in. Even if the occurrence of certain features of a situation were not explicitly provided as input for that specific case, the presence of those features had to have been previously introduced as background knowledge (conjectures or assumptions) without which no such inferences could have been drawn.

An adequate explanatory argument for the occurrence of any explanandum event involves its subsumption by law. When that covering law is deterministic, the explanation involved will be a deterministic-deductive one:

$$\text{(DD)}\quad \begin{array}{l} (x)(t)[(F1xt \,\&\, \ldots \&\, Fmxt) \;{=}u{\Rightarrow}\; Axt^{*}] \\ F1at \,\&\, \ldots \&\, Faxt \\ \hline Aat^{*} \end{array} \quad [u]$$

where "[u]" indicates that the degree of nomic expectability in this case is u. When it is indeterministic, it will be an indeterministic-probabilistic instead:

$$\text{(IP)}\quad \begin{array}{l} (x)(t)[(F1xt \,\&\, \ldots \&\, Fmxt) \;{=}n{\Rightarrow}\; Axt^{*}] \\ F1at \,\&\, \ldots \&\, Faxt \\ \hline\hline Aat^{*} \end{array} \quad [n]$$

where "[n]" indicates that the degree of nomic expectability in this case is n. Adequate explanations of either kind, however, presuppose the satisfiability of the requirement of maximal specificity with respect to their lawlike premises, which otherwise cannot possibly be true, as we discovered in Chapter 5.

To the extent to which SAM is employed as a mode of simulation for the purpose of deriving predictions, it seems to be on a sound theoretical footing within the limitations defined by its character as a species of statistical reasoning. To the extent to which PAM attempts to go beyond the derivation of predictions to providing causal explanations, however, bear in mind that these explanations are adequate only if they are true. A story about John's reason for robbing a store, say, might be subject to analysis by PAM with very plausible results. But those results are in fact "explanatory" only so long as they reflect the relevant factors that affected their subjects: that their subjects may have acted "as if" they had those motives and beliefs is not good enough! Their adequacy therefore rests upon the knowledge base that they represent, which lends further support to the notion that perhaps scripts and frames are best understood as expert systems of a special kind.

SELECTING A REPRESENTATION SCHEME

The choice between different schemes for representing knowledge appears to require coordination between different kinds of epistemic content and different kinds of logical structure. Recall, for example, the conditions of consistency, closure, partial evidence, and complete evidence that have been discussed above. These conditions are normative requirements that define the properties distinctive of rationality of belief. To the extent that they are satisfied by z's beliefs at t, z at t tends to qualify as rational with respect to his beliefs, and, to the extent to which they are not satisfied, as irrational with respect to his beliefs. It must not be overlooked, therefore, that the extent to which they are satisfied in any specific case is a matter that requires empirical determination: it can vary from person to person and time to time. This is a synthetic question that is unanswerable *a priori*.

It follows that the greater the extent to which the beliefs that require representation display rationality of belief, the greater the extent to which those beliefs and habits of thought should correspond to normative expectations. We expect a higher degree of rationality of belief from a domain expert than we do of a personal friend, but we expect a higher degree of rationality of belief from a personal friend than we do of an inmate at a lunatic asylum (assuming, of course, they are not the same). This means that the principles of deductive reasoning and of inductive reasoning are ordinarily exemplified to a higher degree by domain experts than by our friends and by our friends to a higher degree than by inmates of asylums.

While we may expect irrationality in the beliefs of the insane, the existence of intensional contexts should indicate that some degree of violation of rationality of belief ought to be expected of ordinary human beings. Indeed, the three kinds of cases of intensionality that we have explored here already suggest different ways in which these requirements are subject to violation. Consider, for example, cases of kind (II), in which Bob possesses a belief of the form, '$p \rightarrow q$', but fails to believe 'q' even though 'p' happens to be true. Of the three, this case is the least intriguing, no doubt, because it can be explained as an effect of ignorance. If Bob believes '$p \rightarrow q$', but does not believe 'p', we would not expect him to come to believe 'q'

– unless he possesses other evidence supporting 'q', for example, in which case such an intensional context may tacitly reflect a violation of the partial evidence condition (CR-3) or of the complete evidence condition (CR-4), respectively.

Cases of kind (III), in which Bob possesses a belief of the form, '$p \rightarrow q$' and a belief of the form 'p', yet fails to believe 'q' even though 'q' has to be true if his other beliefs are true, are therefore more intriguing. An instance of this kind cannot be explained away as an effect of ignorance, but rather reflects a failure to satisfy the condition of closure (CR-l). Similarly, cases of kind (IV), in which Bob possesses a belief of the form, '$p \rightarrow q$', a belief of the form 'p' and another belief of the form '$-q$', cannot be explained as an effect of ignorance or as a failure of closure. For, in cases of this kind, those who hold belief sets with these properties fail to satisfy the requirement of consistency (CR-2). The difference between the normative and the descriptive, therefore, is clearly on display, since the extent to which these normative constraints happen to be satisfied by human beings – or even by experts with respect to their domains – must not be presupposed.

Levesque (1986a) suggests that the principal difficulty with reasoning is what he calls "incomplete knowledge". Thus, for example, he suggests that,

> A knowledge base (or KB) is called *incomplete* if it tells us which of a number of sentences is true but doesn't tell us which one. So, for example, a KB that tells us Jack is married to either Jan or Jill without telling us which of the two he is married to would be incomplete. [Levesque (1986a), p. 91].

Incomplete knowledge, in this case, arises due to knowledge of the truth of *disjunctions* without knowing which of its disjuncts happens to be true. Indeed, *existential generalizations* (knowing that Jack is married to someone without knowing whom), *negations* (knowing that Jan is not Jack's wife but not knowing who is), and *universal generalizations* (knowing that every computer scientist at the party likes Chinese food but not knowing who they are) are also examples of incomplete knowledge that are identified by Levesque.

However ubiquitous its occurrence may appear to be, however, the existence of incomplete knowledge merely reflects the absence of our knowledge with respect to various domains without affecting the principles of inference (or "the rules of reasoning") that we employ in its pursuit. The patterns that Levesque discerns, such as those involving disjunction, are cases in which an inference could be drawn but for a missing premise. The pattern involved here – namely, if '*p* or *q*' but '-*p*', then infer '*q*' – exemplifies a valid principle of deductive logic, which is not in dispute. In this respect, it represents an intensional context like that represented by (II) rather than an intensional context like that represented by (III) or by (IV). Neither incomplete knowledge nor cases like (II) violate principles of inference in the sense in which cases like (III) and (IV) violate principles of inference. The failure to draw an inference by Elimination (or a "Disjunctive Syllogism") by virtue of ignorance is quite different than having knowledge but violating Modus Ponens.

Logical Structures. A distinction is therefore required between *content* and *reasoning* with respect to any human being. For the logical properties of their domain content (including their sets of beliefs, especially) itself can exhibit such properties as logical consistency and deductive closure, even if the patterns of reasoning that they exemplify are hopelessly defective. One is a product, the other is a process. Thus, the principal advantage of semantic networks, as we have unpacked them, is that their employment for the representation of a set of beliefs imposes no contraints with respect to the consistency or the closure of the content within that domain. Incomplete knowledge is no problem for semantic networks. And they do not assume the applicability of Modus Ponens or of any other valid rules of inference.

The use of production systems, by contrast, clearly implies the applicability of the principles of sentential logic, without which their implemention would be inconceivable. And this, of course, makes a great deal of sense, since inconsistent or incomplete "expert knowledge" would tend to defeat its purposeful existence. Moreover, because sentential logic is a special case of predicate calculus, while scripts and frames are an enhanced version, the conditions imposed by the logical structures that they utilize increase from semantic networks to production systems to predicate calculus right up to scripts and frames. The use of enhanced predicate calculus, of course, does not dictate that those whose behavior is thereby described must always be consistent but rather that a script describing that behavior itself (however unstable it might be across time) satisfies the constraints thereby imposed.

The divergence between a normative conception of knowledge (for which any person must be logically consistent and evidentially justified in holding their beliefs) and a descriptive conception of knowledge (for which persons might hold beliefs that vary from person to person and from time to time, where neither consistency nor evidence matter), therefore, provides a continuum for relating these alternative schemes for representing knowledge:

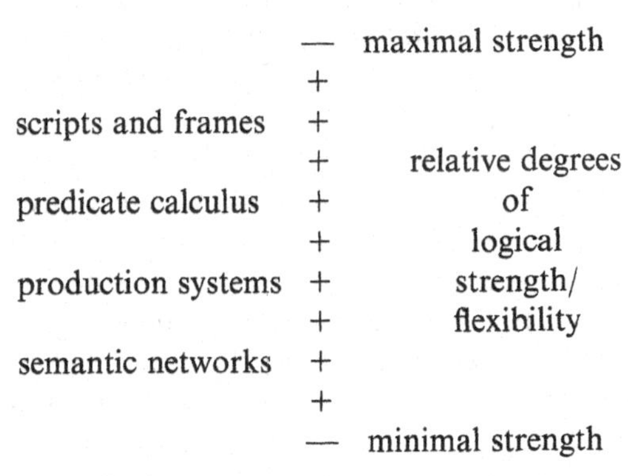

Fig. 33. A Logical Structure Comparison.

where semantic networks reflect the weakest commitments to normative standards of reasoning, whereas scripts and frames reflect the strongest.

There are alternative perspectives on the relationships between what I am taking to be alternative modes of knowledge representation, of course. Production systems, for example, are often viewed as a special type of architecture rather than as knowledge-based programs of a particular kind. "Scripts" and "frames" can be differentiated from one another, moreover, just as "expert systems" and "production systems" can be differentiated from one another. My purpose here, therefore, is not to deny that there are ways in which they can be viewed as different sorts of things but rather to attempt to discover a framework relative to which they might be systematically compared. I shall not take offense if there are those who insist that "category mistakes" are being committed by comparing things of such very different kinds. But I would propose another point of view.

Epistemic Contents. If these are the logical structures that are available for the purpose of representing knowledge, then when would it be appropriate to use them? The principal purpose for which semantic networks, production systems, predicate calculus, and the like stand as alternatives is representing the kinds of content and capturing the kinds of relations that are relevant to human knowledge and human beliefs. These include representing the information that may be found within the human mind, especially, say, the organization of a person's memory, the structure of his knowledge, and so on, where distinctions between descriptive and normative approaches to matters such as these tend to be ignored. Discovering the answer to this question, however, requires its deliberate consideration. [Cf. Brachman and Smith (1980), for various perspectives on these issues.]

A descriptive framework is intended to be used to describe how various persons' memories are organized or how their knowledge is structured, etc. But there are good memories and poor ones; some have photographic memories and others suffer from Alzheimer's disease. Moreover, since people are typically inconsistent in their beliefs, it would be a mistake to employ an approach that imposes consistency upon them – if the intended effect is to describe what they do believe. The beliefs that are possessed by human beings do not always qualify as even minimally rational. Since sets of inconsistent beliefs cannot all be true, however, it would *not* be a mistake to utilize an approach that imposes consistency upon them – if the intended effect is prescriptive. When we aim at expert knowledge, after all, we are searching for beliefs that are maximally rational with respect to a specific knowledge domain. Beliefs that merely satisfy consistency and closure are simply not enough. We want beliefs that are supported and hopefully true.

There thus appears to be a progression in relation to the conditions we expect of different kinds of knowledge or belief. The beliefs of ordinary human beings, no doubt, may be inconsistent or incomplete, and partially or entirely unjustified. With respect to domain experts, however, we want

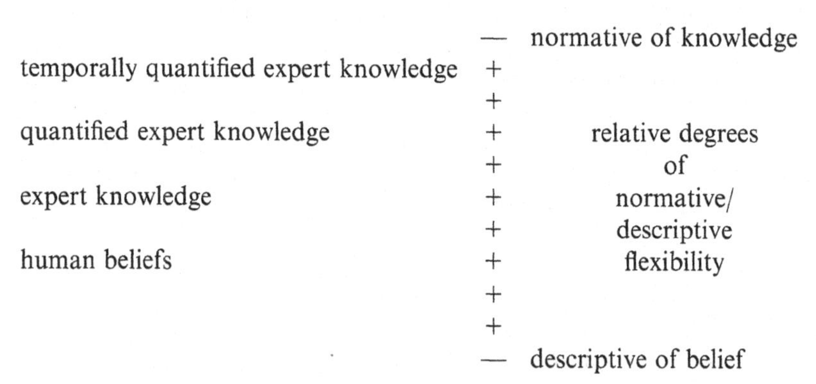

Fig. 34. A Comparison of Epistemic Contents.

consistency and closure for sets of beliefs that satisfy appropriate requirements for evidence and warrants. When that expert knowledge involves quantification, moreover, we expect more of the same. When it extends to temporally quantified knowledge domains, we expect the most. Thus, successively more stringent conditions are satisfied by these very cases, as the comparison of epistemic content in Fig. 34 is intended to display.

The greater the emphasis upon expertise with respect to a specific domain, the greater the extent to which normative constraints should be satisfied; conversely, the less the emphasis upon expertise with respect to a specific domain, the less the extent to which normative constraints matter after all.

It therefore appears as though there should be a suitable degree of correspondence between the epistemic content to be represented with regard to choice of domain, level of analysis, and source of information, on the one hand, and the logical structure that would be appropriate to represent that content with regard to choice of vocabulary, formation rules, and transformation rules, on the other. And, indeed, that does appear to be the case. For the representation of the beliefs of ordinary human beings (complete with faulty memories and logical inconsistencies) requires a scheme of representation that does not presume even minimal rationality, which semantic networks afford. While semantic networks can be employed in a manner that satisfies stronger constraints, they alone do not impose them as "the rule".

The representation of expert knowledge (in relation to some domain of expertise), by comparison, requires a scheme of representation that implements conditions of consistency and closure, which can be satisfied by production systems, by predicate calculus, and by scripts and frames alike. A choice between these alternatives, therefore, would appear to depend upon the precise character of that knowledge, where the least elaborate scheme that succeeds in its representation would be an appropriate choice. Thus, when that knowledge can be adequately formalized by means of sentence logic, then production systems fill the bill; when that knowledge requires quantification, predicate calculus can be employed; and when that knowledge requires temporal quantification as well, scripts and frames are available. The situations in which these different schemes have actually been employed, therefore, tend to satisfy the recommendations of this account.

HAS THE ROLE OF LOGIC BEEN MISCONCEIVED?

There are those, however, who might object to this entire line of analysis, precisely because of the prominent role that has been played by logic. Each of these schemes of knowledge representation, after all, has been portrayed as exemplifying some specific species of logical form. For semantic networks have been associated with sentential functions, production systems with sentential logic, predicate calculus with predicate calculus, and scripts and frames with temporally quantified predicate calculus. If there are reasons for believing that logic has been miscast in its role within this scheme, therefore, then this account itself might have been misconceived. Indeed, there are those who appear to doubt that logic could possibly deserve such a fundamental role with respect to the foundations of knowledge representation on grounds that we are now in a position to review.

The foremost opponent of "the logistic approach" within artificial intelligence is Marvin Minsky, who has sought to discredit the employment of logic for several different reasons. Minsky has indicated what he takes to be various factors that inhibit the use of logic for the purpose of representation, including difficulties in formalizing knowledge bases, especially if they are intended to capture common-sense knowledge; the necessity to identify what is relevant knowledge and what is not, especially with respect to the frame problem; the restriction of logic to monotonic systems, wherein an inference, once drawn, can never be rejected; the difficulties in capturing the role of control in managing the production of deductions; and the combinatorial explosions encountered in developing and managing knowledge bases that are rich and detailed [Minksy (1975), pp. 123-126].

The criticism that he appears to take most seriously, however, reflects his concern over the imposition of conditions of consistency and completeness. Thus, Minsky observes that ordinary beliefs are often not consistent:

> Consistency ... requires one's axioms [sets of beliefs] to imply no contradictions. But I do not believe that consistency is necessary or even desirable in a developing intelligent system. No one is ever completely consistent. What is important is how one handles paradox or conflict, how one learns from mistakes, how one turns aside from suspected inconsistencies. [Minsky (1975), p. 126]

For reasons such as these, Minsky arrives at such conclusions as (a) that logical reasoning is not sufficiently flexible to serve as a foundation for thinking, (b) that the consistency that logic "absolutely demands" is not ordinarily available and is probably not even desirable, (c) that general rules of inference should not be separated from the specific knowledge that a knowledge base provides, and the like [cf. Minsky (1975), p. 127].

Because particular aspects of Minsky's position are well-founded and true, it may be hard to appreciate that much of the spirit of his position is misleading and false. It is difficult to formalize common-sense knowledge; there are real problems about relevance; the role of control poses genuine difficulties; and combinatorial explosions are not easy to avoid. Since monotonic reasoning – for which any conclusion that follows from a specific set of premises continues to follow from any superset including those premises – is deductive reasoning, it would be very difficult, if not impossible, to implement the dynamics of data-bases that change across time with the addition of new data, were monotonic reasoning the only kind of reasoning that happened to be available for our use.

Deductive reasoning is not the only kind of reasoning available for our use, of course, and the implementation of data-bases of this kind depends on the use of reasoning that is inductive and non-monotonic. Minsky's position on consistency and completeness, moreover, seems to trade upon an equivocation. Once a distinction is drawn between:

(D15)	the context of discovery	=df	the invention of hypotheses and theories, apart from any concern for their origin or for their form, and
(D16)	the context of justification	=df	the evaluation of hypotheses and theories on the available evidence, in light of the rules of deduction and of induction,

it should be evident that, while "dreaming up" hypotheses and theories *is* a kind of thinking that ought to take place independently of concern for completeness and consistency, the process of their rational appraisal as warranted for acceptance as true beliefs is *not* [Hempel (1966), Ch. 2].

If we identify "thinking" with the context of discovery and "reasoning" with the context of justification, then it is indeed correct that logical reasoning is not sufficiently flexible to serve as a foundation for *thinking*. But it certainly does not follow that logical consistency and deductive closure are therefore properties that are neither available nor desirable within contexts of *reasoning* in general. So long as the sets of beliefs that schemes are intended to represent are beliefs of ordinary human beings, even minimal rationality may be too strong a requirement. But in relation to expert knowledge, it is too weak.

Thus, with respect to thinking, there are no reasons to doubt that human beings are almost never entirely consistent. And there are no reasons to doubt that how human beings handle paradox or conflict, how they learn from mistakes and turn away from suspected inconsistencies are crucial aspects of thought. But things are different with reasoning. An expert whose knowledge could not possibly be true because it was inconsistent would not be worth the bother. An expert whose knowledge was consistent but unsupported by the available evidence would hardly be worth utilizing. Perhaps the underlying moral that emerges from this discussion, therefore, is that the most important decisions confronting those working in this field involve determining exactly what "knowledge" is worth representing.

8. PROGRAM VERIFICATION

A potentially valuable distinction that might capture at least part of the difference between experts and expert systems has been drawn by Drew McDermott (1981b). McDermott suggests that there is a fundamental difference between *finished programs* and *expert systems*. Thus, in explaining the development of the expert system R1, which is employed to configure VAX-11 computer systems, he expresses concern for a rather subtle issue:

> It is not clear that all (or even most) of R1's supporters realize that R1 will always make mistakes. The problem is that at least some of R1's supporters think of it as a program rather than as an expert. There is, of course, a big difference between programs and experts. Finished programs, by definition, have no bugs. When experts are finished, on the other hand, they're dead. [McDermott (1981b), p. 29]

Since an expert system reflects an expert's knowledge represented in the form of a computer program, however, the juxtaposition of "experts" and "programs" is somewhat misleading, insofar as "experts" in this sense *are* computer programs. But McDermott may yet have a point, because eventhough experts are programs, they are programs *of a certain special kind.*

Several issues are involved here, insofar as the type of knowledge that an expert system reflects might differ in its kind from that represented by other programs, such as those that can perform purely mathematical manipulations of addition, subtraction, and the like. One difference appears to be that the inference rules for mathematical procedures are purely deductive, in the sense that they follow as analytic consequences from a theory of numbers as *a priori* knowledge. The entities and relations that they are intended to reflect exist in abstract domains. The typical expert system, by contrast, cannot possibly rely exclusively upon deductive rules of inference, since the knowledge that they represent includes synthetic consequences based upon some body of *a posteriori* knowledge. The entities and relations which they are intended to reflect exist in physical domains. They are fallible theories.

McDermott's suggestion that R1 will always make mistakes, of course, is especially interesting in view of the signifiance attached to the capacity to make mistakes as it was elaborated in Chapter 2. But it ought to be borne in mind that, when the results of a program are not what we wanted them to be, the "mistakes" that are involved may be those of the users of the system rather than those of the system itself. The point of this remark, as McDermott may have meant for it to be understood, could be that systems of this kind function on the basis of crude generalizations and "rules of thumb" for which exceptions are inevitable. Whenever there is more variation in the phenomena than can be programmed into the machine, there will tend to be situations or arrangements that are physically possible in relation to the world which are not encompassed within the scope of such a program.

It would be a wonderful thing, of course, if the performance of any program could be anticipated with the certainty that is characteristic of deductive domains. A conception of this kind has been proposed by C. A. R. Hoare, who maintains that programming should strive to be more like mathematics:

> I hold the opinion that the construction of computer programs is a mathematical activity like the solution of differential equations, that programs can be derived from their specifications through mathematical insight, calculation, and proof, using algebraic laws as simple and elegant as those of elementary arithmetic. [Hoare (1986), p. 115]

This approach, no doubt, is extraordinarily appealing. For Hoare holds that verification procedures promise the prospect of a generally applicable and completely reliable method for guaranteeing the performance of a program. Even when the knowledge that programs represent cannot be established *a priori*, at least the properties of programs can still be known with certainty.

Others, such as Richard DeMillo, Richard Lipton, and Alan Perlis (1979), contend that any such position rests upon a major misconception and thus appears to be mistaken. They suggest that social processes play an important role in accepting the validity of a proof or the truth of a theorem, no matter whether within purely mathematical contexts or without: "We believe that, in the end, it is a social process that determines whether mathematicians feel confident about a theorem" [DeMillo, Lipton, and Perlis (1979), p. 271]. Indeed, the situation with respect to the verification of programs is even worse, because similar social processes do not take place between program verifiers. The use of verification to guarantee the performance of a program is therefore bound to fail. While these authors do not focus upon Hoare's work in their discussion, there is no reason to doubt that his account and that of others, such as E. W. Dijkstra (1976), who share a similar point of view, is the methodological position meant to bear the brunt of this criticism.

Their contribution has aroused enormous interest and considerable controversy, ranging from unqualified agreement expressed, for example, by Daniel Glazer ["Such an article makes me delight in being ... a member of the human race" (1979), p. 621] to unqualified disagreement expressed, for example, by W. D. Maurer ["The catalog of criticisms of the idea of proving a program correct ... demands a catalog of responses ..." (1979), p. 625]. Indeed, some of the most interesting reactions have come from those whose position lies somewhere in between, such as J. van den Bos, who maintains that, "Once one accepts the quasi-empiricism in mathematics, and by analogy in computer science, one can either become an adherent of the Popperian school of conjectures (theories) and refutations (1965), or one may believe Kuhn (1970), who claims that the fate of scientific theories is decided by a social forum" [van den Bos (1979), p. 623]. [Some fascinating articles on Popper's and Kuhn's similarities and differences can be found in Lakatos and Musgrave (1970).]

Perhaps better than any other commentator, van den Bos seems to have put his finger on what may very well be the underlying issue raised by DeMillo, Lipton, and Perlis, namely: if program verification, like mathematical validation, could only occur as the result of a fallible social process, if it could occur at all, then what would distinguish programming procedures from other evaluative activities like judges deciding cases at law and referees reviewing articles for journals? If it is naive to presume that mathematical demonstrations, program verifications, and the like, are fundamentally dissimilar from these eminently fallible activities, on what basis can they be distinguished? Is the correctness of a program, like the truth of a theorem or the validity of a proof, merely determined by the subjective judgment of its programmer?

The issues that are involved here are fundamental to understanding the similarities and the differences between mathematics and programming, on the one hand, and algorithms and programs, on the other. They raise major conceptual and theoretical problems for understanding the relationship between the program and the machine, especially with respect to the extent to which we can establish *a priori* expectations concerning a machine's behavior. Even though these difficulties are not peculiar to artificial intelligence *per se*, they are extremely relevant to grasping methodological and epistemological dimensions of the computer science enterprise within which artificial intelligence assumes signifiance. To illuminate our epistemic and ethical relations to these machines, therefore, this chapter will be devoted to their exploration.

THE COMPARISON WITH MATHEMATICS

DeMillo, Lipton, and Perlis defend their position by drawing a distinction between "proofs" and "demonstrations", where demonstrations are supposed to be long chains of formal logic, while proofs are not. The difference which may be intended here, strictly speaking, appears to be that between "proofs" and what are typically referred to as "proof sketches", where proof sketches are incomplete (or "partial") proofs. Indeed, "proofs" are normally defined in terms of demonstrations, where a proof of theorem *T* occurs just in case the theorem *T* has been shown to be the last member of a sequence of formulae where every member of that sequence is either given (as an axiom or as an assumption) or else derived from preceding members of that sequence (by relying on the members of a specific set of inference rules) [Church (1959b), p. 182]. What is called "mathematical induction" is a special case of deductive procedures applied to infinite sequences, even though its name tends to obscure the demonstrative nature of these techniques [Black (1967), p.169].

Moreover, when DeMillo, Lipton, and Perlis are interpreted as taking "proof sketches" rather than "proofs" as the objects of mathematicians' attention, it becomes possible to make good sense of otherwise puzzling remarks such as:

> In mathematics, the aim is to increase one's confidence in the correctness of a theorem, and it's true that one of the devices mathematicians could in theory use to achieve this goal is a long chain of formal logic. But in fact they don't. What they use is a proof, a very different animal. Nor does the proof settle the matter; contrary to what its name suggests, a proof is only one step in the direction of confidence. [DeMillo, Lipton, and Perlis (1979), p. 271]

Hence, while a "proof", strictly speaking, is a (not necessarily long) chain of formal logic that is no different than a demonstration, a "proof sketch" is "a very different animal", where a proof sketch, unlike a proof, may often be "only one step in the direction of confidence". Nevertheless, although these reflections offer an interpretation under which their sentences appear to be true, it leaves open a larger question, namely: whether the aim of proofs in mathematics can be adequately envisioned as that of "increasing one's confidence in the correctness of theorems" rather than as formal demonstrations.

In support of this depiction, DeMillo, Lipton, and Perlis emphasize the tentative and fallible character of mathematical progress, where out of some 200,000 theorems said to be published each year, "A number of these are subsequently contradicted or otherwise disallowed, others are thrown into doubt and most are ignored". Since numerous purported "proofs" are unable to withstand criticial scrutiny, they suggest, the acceptability or believability of a specific mathematical result depends upon its reception and ultimate evaluation by the mathematical community. In this spirit, they describe what appears to be a typical sequence of activity within this arena, where, say, a proof begins as an idea in someone's mind, receives translation into a sketch, is discussed with colleagues, and, if no substantial objections are raised, is developed and submitted for publication. Then, if it survives the criticism of fellow mathematicians, it tends to be accepted as a valid demonstration.

In this sense, the behavior of typical members of the mathematical community in the discovery and promotion of specific findings certainly assumes the dimensions of a social process involving more than one person interacting together to bring about a certain outcome. Although it may be difficult to imagine a Bertrand Russell or a David Hilbert rushing to his colleagues for their approval of his findings, there would appear to be no good reasons to doubt that, generally speaking, average mathematicians frequently behave in the manner that has been described. In this sense, I would tend to agree that mathematicians' mistakes are typically discovered or corrected as a result of informal interactions with other mathematicians. The restraints imposed by symbolic logic, after all, exert their influence only through their assimilation as habits of thought and as patterns of reasoning by specific members of a community of this kind: discoveries and corrections of mistakes often do occur when one mathematician nudges another "in the right direction". [Relevant discussions can be found in Bochner (1966) and in Lakatos (1976).]

To whatever extent DeMillo, Lipton, and Perlis should be regarded as endorsing the view that review procedures exercised by colleagues and peers tend to improve the quality of papers that appear in mathematics journals, there seems to be little ground for disagreement. For potential proofs are often strengthened, theorems altered to correspond to what is provable, and various arguments discovered to be flawed through social interaction. Nonetheless, a community of mathematicians who are fast and sloppy referees is no more difficult to imagine than a community of quack physicians. Where, for example, is the university whose faculty does not occasionally compose somewhat shoddy and unreliable reviews, even for journals of exceptional standing? What makes what we call "a proof" a proof, after all, is its validity rather than its acceptance by us as valid, just as what makes a sentence true is that what it asserts to be the case is the case, not merely that it is believed by us and therefore referred to as "true". An invalid proof can be mistaken as valid, just as a valid proof can be mistaken as invalid. Strictly speaking, therefore, social processing is neither necessary nor sufficient for any proof to be valid, as DeMillo, Lipton, and Perlis [(1979), p. 272] implicitly concede.

Confidence in the truth of a theorem or in the validity of an argument, of course, appears to be a psychological property of a person-at-a-time. One and the same person at two different times can vary greatly in confidence over the truth of the same theorem or the validity of the same argument, just as two different persons at the same time might vary greatly in their confidence that that same theorem is true or that that same argument is valid. Indeed, there is nothing inconsistent about scenarios in which, say, someone is completely confident that a specific formula is a theorem when it happens to be false, or else completely uncertain whether a particular argument is valid when it happens to be valid. No doubt, mathematicians are sometimes driven to discover demonstrations of theorems after they are already completely convinced of their truth: demonstrations, in such cases, cannot increase the degree of confidence when that degree is already maximally strong. But they might still fulfill other (non-psychological) purposes by providing objective evidence of the truth of one's subjective belief.

From the perspective of the theory of knowledge, the role of demonstrations is readily apparent. The standard conception of knowledge, after all, implies that an individual z who is in a state of belief with respect to some specific formula f (when z believes that f is a theorem) cannot be properly qualified as possessing knowledge that f is a theorem unless that belief can be supported by means of reasons, evidence, or warrants, which, as we previously discovered, might be one or another of three very different kinds, depending upon the nature of the objects that might be known. For results in logic and mathematics fall within the domain of deductive methodology and require demonstrations. Lawful and causal claims, however, fall within the domain of empirical inquiries and require inductive warrants. And observational and experimental findings fall within the domain of perceptual investigations and acquire support on the basis of sense experience.

Concepts of Verification. With respect to demonstrations, the term "verification" can be used in two rather different ways. The first occurs in pure mathematics and pure logic, in which "theorems" of mathematics and logic are subject to derivation. These theorems characterize claims that are always true as a function of the meanings assigned to the particular symbols by means of which they are expressed. When they are true, therefore, they are analytic (where their truth can be established *a priori*). Theorem schemata and theorems are subject to "verification" in this sense by deriving them from no premises at all (within systems of natural deduction) or from primitive axioms (within axiomatic formal systems). [A lucid discussion of the differences between natural deduction systems and axiomatic formal systems, where the former operate with sets of inference rules in lieu of formal axioms, but the latter operates with sets of axioms and as few as a single rule, moreover, can be found in Blumberg (1967).]

The second occurs in ordinary reasoning and in scientific contexts in general, whenever "conclusions" are shown to follow from specific sets of "premises" where there is no presumption that these conclusions might be derived from no premises at all or that these conclusions should be true as a function of their meaning. When they are true, they are synthetic (and their truth can be established only *a posteriori*). Thus, within a system of natural deduction or an axiomatic formal system, the members of a class of consequences that can be derived from no premises at all or that follow from primitive axioms alone should be described as "absolutely verifiable". But the members of a class of consequences that instead can only be derived in relation to specific sets of premises whose truth is not absolutely verifiable, by comparison, ought to be described as "relatively verifiable".

The difference between "absolute" and "relative" verifiability, moreover, is extremely important for the theory of knowledge. For theorems that can be "verified" in the absolute sense cannot be false (so long as the rules are not changed or the axioms are not altered). But conclusions that are "verified" in the relative sense can still be false (even when the premises and the rules remain the same). The absolute verification of a mathematical theorem thus satisfies both necessary and sufficient conditions for its warranted acceptance as true, but the relative verification of a conclusion

– even in mathematics – does not. Indeed, as an epistemic policy, the degree of confidence that anyone should invest in the conclusion of an argument should never exceed the degree of confidence that ought to be invested in its premises – even when it is valid! Unless the premises of an argument cannot be false – unless those premises themselves are absolutely verifiable – it is a mistake to assume that the conclusion of a valid argument cannot be false. No more can appropriately be claimed than that its conclusion must be true "if" its premises are true, which is the principal definitional property of a deductive argument that happens to be valid.

The truth of the conclusion of a valid deductive argument, therefore, can never be more certain than the truth of its premises, unless its truth can be established on independent grounds. While deductive reasoning preserves the truth (in the sense that the conclusion of a valid argument cannot be false when its premises are true), the truth of those premises can be guaranteed, in general, only under the special circumstances that arise when they themselves are verifiable in the absolute sense. Otherwise, the truth of the premises of any argument has to be established on independent grounds, which might be deductive, inductive, or perceptual. Yet none of these types of warrants provides an infallible foundation for any inference to the truth of the conclusions that they happen to support – apart from those deductive arguments which are absolutely verifiable.

Perhaps few would be inclined to think that our senses are infallible, since that would only be the case if things must always be the way they appear to be. The occurrence of illusions, hallucinations, and delusions disabuses us of that particular fantasy. The mistakes we make in thinking about modes of inference are more likely to occur regarding inductive and deductive reasoning, whose features are often not well understood. We have discovered that valid deductive arguments are demonstrative, nonampliative, and additive, while proper inductive arguments are nondemonstrative, ampliative, and additive. Arguments of these two kinds are alike insofar as they involve inferences from premises in the form of sentences to conclusions in the form of sentences, where an argument is acceptable when and only when it satisfies acceptable rules of inference.

Perceptual inference, by contrast, entails the application of language for the purpose of describing the contents of more or less direct experience, where no special premises are involved (over and beyond the use of some specific language, *L*, which deductive and inductive arguments require as well). The relationship between the premises and the conclusion of deductive and inductive arguments, therefore, has the character of a logical relation between specific forms of language, while the utilization of language for perceptual inference has the character of a causal interaction between a person and the world, as Chapter 4 has explained. But we should not overlook that, apart from imagination and conjecture, which serve as sources of ideas but do not establish their acceptability, our states of knowledge other than those of pure mathematics and logic ultimately depend for their support upon their connections to experience.

It ought to be noticed that the differences between deductive and inductive arguments that have been presented are defined on the basis of semantical properties, but the definition of "proof" provided earlier was based upon syntactical properties. To see that this is the case, recall that semantics encompasses whatever means whereby one thing might stand for another, subsuming the concepts of truth, meaning, and interpretation. If an argument is *demonstrative* when the falsity of its conclusion would be inconsistent with the truth of its premises, *ampliative* when its conclusion contains more content than do its premises, and so on, then these are semantic conceptions which, as it happens, possess syntactic counterparts. An argument that is demonstrative may be characterized syntactically as a "proof", an argument that is additive could be described as "monotonic", and so on. One reason for the importance of social processes in the production of proofs in mathematics – and for the importance of its absence from the production of verifications in computer science – therefore might be the opportunity it avails for more than one mathematician to inspect it for a suitable relationship linking its syntax and its presumed semantics.

Because computer programs, like mathematical proofs, are syntactical entities consisting of sequences of lines (strings of characters, and the like), both appear to be completely formalized objects for which completely formal procedures appear to be appropriate. Yet programs differ from theorems, at least to the extent to which programs are supposed to possess a semantic significance that theorems seem to lack. For the sequences of lines that compose a program are intended to stand for operations and procedures that can be performed by a machine, whereas the sequences of lines that constitute a proof are not. Even if the social acceptability of a mathematical proof is neither necessary nor sufficient for its validity, therefore, the suggestion might be made that the existence of social processes of program verification could be even more important for the success of this endeavor than it is for verifying proofs. And the reason that could be advanced in support of this position is the opportunity that such practices provide for more than one programmer to inspect a program for a suitable relationship beween its syntax and its intended semantics, i.e., the behavior that is expected of the machine.

It could be argued that even purely mathematical proofs may be viewed as possessing semantical relations to some domain of abstract entities, which might be variously construed as Platonism, Intuitionism, or Conventionalism take them [cf. Benacerraf and Putnam (1963)]. Similarly, in the case of programming formulae, they might be interpreted as data structures of various kinds that stand for various entities and relations within either abstract or physical domains (including, but not restricted to, those that occur in purely physical domains). But even if that is true, the existence of an abstract domain for mathematical proofs or for computer programs would not alter the differences resulting from the existence of a domain of physical entities that exists in the case of programs but does not exist in the case of proofs: a program bears a relation to the behavior of a machine that is absent for a proof.

DeMillo, Lipton, and Perlis are inclined to emphasize their finding that the verification of programs, unlike the construction of proofs, is not an especially popular pastime. The argument we are considering tends to reinforce the importance of this disparity in behavior between the members of the mathematical community and the members of the programming fraternity. Their position implicitly presumes that programmers are inherently inhibited from collaborating on the verification of programs, especially because it is such a tedious and complex activity. Suppose, however, that incentives and other rewards in our society were transformed in some direct and obvious ways, where substantial financial benefits were offered for the best team efforts in verifying programs, say, with prizes of up to $10,000,000 awarded by Ed McMahon and guest appearances on *The Tonight Show*. Surely under conditions of this kind, the past tendency of program verifiers to engage in solitary endeavors might be readily affected, resulting in the emergence of a new wave in program verification – "the collaborative verification group" – dedicated to mutual efforts by more than one programmer to verify particular programs. Under these circumstances, which are not completely far-fetched within the context of modern times, social processes for the verification of programs in the computer science community could emerge that would be counterparts to current social processes for the validation of proofs in mathematics. And were this to occur, they would no longer display any difference of this kind.

If this were to come about, then the primary assumption underlying the position DeMillo, Lipton, and Perlis defend would no longer apply. Regardless of what other differences might distinguish them, in this respect their social processing – of theorems and of programs – would be the same. Let us refer to differences between subjects or activities that cannot be overcome no matter what efforts we might undertake as differences *in principle* and refer to those that can be overcome by making appropriate efforts as differences *in practice*. Then it should be obvious that DeMillo, Lipton, and Perlis have identified a difference in practice that is not also a difference in principle. What this means is that, to the extent to which their position depends upon this difference, it represents no more than a contingent, *de facto* state of affairs that might be merely a transient stage in the development of program verification within the computer science community. If there is an in principle difference between them, therefore, it must lie elsewhere, because this divergence in social practice is a difference that could thus be overcome.

Probable Verifications. The function of a program, no doubt, is to convey instructions to a computer. Most programs are written in high-level languages (such as Pascal, Lisp, and Prolog), which simulate "abstract machines" whose instructions can be more readily composed than can those of the machines that execute them. Thus, the *source program* written in a high-level language is translated into an equivalent low-level *object program* written in machine language (by an interpreter as it is written, line by line; or by a compiler after it is written, program by program). The "target machine" can then execute the object program when the system's user instructs it to do so.

In order for programs to be verifiable, they have to be subjectable to deductive procedures. Indeed, precisely this conception is endorsed by Hoare:

> Computer programming is an exact science in that all the properties of a program and all the consequences of executing it can, in principle, be found out from the text of the program itself by means of purely deductive reasoning. [Hoare (1969), p. 576]

Thus, if they are absolutely verifiable, then there must exist some "program rules of inference" or "primitive program axioms" permitting inferences to be drawn concerning the performance that a machine will display when such a program is executed. And if they are relatively verifiable, then there must be sets of premises concerning the text of such a program from which it is possible to derive conclusions concerning the performance that the machine will display when it is executed. If these requirements cannot be satisfied, however, then the very idea of "program verification" must be misconceived.

When DeMillo, Lipton, and Perlis [(1979), p. 273] endorse the position that "a proof can, at best, only probably express truth", therefore, it is important to discover exactly what they mean, since "proofs" are deductive and accordingly enjoy the virtues of demonstrations. There are numerous alternatives. In the first place, this claim might reflect the differences that obtain between "proofs" and "proof sketches", insofar as incomplete or partial proofs do not tend to satisfy the same objective standards and therefore need not convey the same subjective certainty as do complete proofs. In the second place, it might reflect the fallibility of acceptance of the premises of such an argument – which, after all, might be valid yet have at least one false premise and thus be unsound – because of which acceptance of its conclusion should be tempered with uncertainty as well. Although both of these ideas find expression in their article, their principal contention appears to be of a different character.

DeMillo, Lipton and Perlis [(1979), p. 273] distinguish between the "classicists" and the "probabilists", where the classicists maintain the position that

> when one believes mathematical statement *A*, one believes that in principle there is a correct, formal, valid, step by step, syntactically checkable deduction leading to *A* in a suitable logical calculus

which is the complete proof lying behind a "proof sketch" as the object of acceptance or of belief. Probabilists, by comparison, maintain the position that

> since any very long proof can at best be viewed as only probably correct, why not state theorems probabilistically and give probabilistic proofs? The probabilistic proof may have the dual advantage of being technically easier than the classical, bivalent one, and may allow mathematicians to isolate the critical ideas that give rise to uncertainty in traditional, binary proofs.

Thus, in application to "proofs" or to "proof sketches", there appear to be three elements to this position: first, that long proofs are difficult to follow; second, that probabilistic proofs are easier to follow; and, third, that probabilistic proofs may help disclose the problematic aspects of ordinary proofs. Rabin's algorithm for testing probable prime numbers is offered as an illustration: "if you are willing to settle for a very good probability that N is prime (or not prime), then you can get it within a reasonable amount oftime – and with [a] vanishingly small probability of error" [(1979), p. 273].

Their reference to "traditional, binary proofs" is important insofar as traditional proofs (in the classical sense) are supposed to be either valid or invalid: there is nothing "probable" about them. I therefore take DeMillo, Lipton, and Perlis to be endorsing an alternative conception, according to which arguments are amenable to various measures of strength (or corresponding "degrees of conviction"), which might be represented by, say, some numeral between 0 and 1 inclusively, with some of the properties that are associated with probabilities, likelihoods, and so on [Fetzer (1981), Part III]. A hypothetical scale could be constructed accordingly such that,

```
—  demonstrative (maximally strong)
+  1.0
+
+
+        various degrees
+  0.5          of
+        partial support
+
+
+  0.0
—

   fallacious (hopelessly weak)
```

Fig. 35. A Measure of Evidential Support.

where degrees of evidential support of measures 0 and 1 are distinguished from worthless fallacies (whose premises, for example, might be completely irrelevant to their conclusions), on the one hand, and from demonstrative arguments (the truth of whose premises guarantees the truth of their conclusions), one the other, as extreme cases not representing *partial* support.

From this point of view, the existence of (even) a "vanishingly small" probability of error is essential to a probabilistic proof, since if there were no probability of error at all, a "proof" could not be "probabilistic".

Thus, if the commission of an error were either a necessity (as in the case of a fallacy of irrelevance) or an impossibility (as in the case of a valid demonstration), then an argument would have to be other than probabilistic. Indeed, the existence of fallacies of irrelevance exemplifies an important point discussed above: fallacious arguments, although logically flawed, can exert enormous persuasive appeal – otherwise, we would not have to learn how to detect and avoid them [Michalos (1969), Ch. 10]. But, if this is the case, then DeMillo, Lipton, and Perlis, when appropriately understood, are advocating a conception of mathematics according to which (i) classical proofs are practically impossible (so that demonstrations are not ordinarily available), yet (ii) worthless fallacies are still unacceptable (so that our conclusions are nevertheless supported). This position strongly suggests the possibility that DeMillo, Lipton, and Perlis should be viewed as advocating the conception of mathematics as within the domain of inductive methodology.

When consideration is given to the distinguishing characteristics of inductive arguments, this interpretation seems to fit their position quite well. Recall, after all, that inductive arguments are (a) non-demonstrative, (b) ampliative, and (c) non-additive. This means (a) that their conclusions can be false even when their premises are true (permitting the possibility of error), (b) that their conclusions contain some information or content not already contained in their premises (otherwise they would not be non-demonstrative), and (c) the addition of further evidence in the form of additional premises can either strengthen or weaken these arguments (whether that evidence is discovered days, months, years, or even centuries later). Thus, in accepting the primality of some very large number N probabilistically, for example, one goes beyond the content contained in the premises (which do not guarantee the truth of that conclusion) and runs a risk of error (which cannot be avoided with probabilistic reasoning). Yet one thereby possesses evidence in support of the truth of such a conclusion, so that its acceptance is warranted, although not conclusively, at least to some degree – which, after all, might be as well as we are able to do.

Indeed, whether or not we can do better appears to be at the heart of the controversy surrounding this position. DeMillo, Lipton, and Perlis, in particular, do not explicitly deny the existence of classical bivalent proofs (although they recommend probabilistic proofs as more appropriate to their subject matter). Moreover, they implicitly concede the existence of classical bivalent proofs (insofar as the pursuit of probabilistic proofs may even lead to their discovery). They even go so far as to suggest that the social processing of a probabilistic proof, like that of a traditional proof, can involve "enough internalization, enough transformation, enough generalization, enough use" that the mathematical community accepts it as correct: "The theorem is thought to be true in the classical sense – that is, in the sense that it could be demonstrated by formal, deductive logic, although for almost all theorems no such deduction ever took place or ever will" [DeMillo, Lipton, and Perlis (1979), p. 274].

The force of their position, therefore, appears to emanate from the complexity that confronts those who would attempt to undertake mathematical proofs and program verifications. They report that a formal demonstration of a conjecture by Ramanujan would require 2000 pages to formalize, and they lament that Russell and Whitehead "in three enormous, taxing volumes, (failed) to get beyond the elementary facts of arithmetic". These specific examples may be subject to dispute. For Ramanujan's conjecture, precisely how is such a fanciful estimate supposed to have been derived and verified? For Russell and Whitehead, is *Principia Mathematica* therefore supposed to be a failure? Yet their basic point ("The lower bounds on the length of formal demonstrations for mathematical theorems are [or, at the very least, certainly can be] immense and there is no reason to believe that such demonstrations for programs would be any shorter or cleaner – quite the contrary" [(1979), p. 276], nonetheless merits our serious contemplation.

One of the most important ambiguities that arises within this context emerges at this juncture; for, while DeMillo, Lipton and Perlis [(1979), p. 278] suggest that the "scaling-up argument" – the contention that very complex programs and proofs can be broken down into much simpler programs and proofs for the purposes of verification and of demonstration – is the best that the other side can produce, they want to deny that it should be taken seriously since it hangs upon an untenable assumption:

> The scaling-up argument seems to be based on the fuzzy notion that the world of programming is like the world of Newtonian physics – made up of smooth, continuous functions. But, in fact, programs are jagged and full of holes and caverns. Every programmer knows that altering a line or sometimes even a bit can utterly destroy a program or mutilate it in ways that we do not understand and cannot predict. [DeMillo, Lipton, and Perlis (1979), p. 278]

Indeed, since this argument is supposed to be the best argument in defense of the verificationist position, they take "the discontinuous nature of programming" as ultimately sounding "the death nell for verification".

Maurer (1979) has objected strenuously to the claims that the verification of one program can never be transferred to any other program ("even a program only one single line different from the original") and that there are no grounds for thinking that the verification of a large program could be broken down into smaller parts ("there is no reason to believe that a big verification can be the sum of many small verifications") [DeMillo, Lipton, and Perlis (1979), p. 278]. In response, he has observed that, while the modification of a correct program can produce an incorrect one for which "no amount of verification can prove it correct" and while minute changes in correct programs can produce "wildly erratic behavior ... if only a single bit is changed", that does not affect the crucial result, namely: that proofs of the correctness of a program can be transferred to other programs when those other programs are carefully controlled modifications of the original program, at least in part if not in whole. Indeed, quite frequently "if a program is broken up into a main program and n subroutines, we have $n + 1$ verifications to do, and that is all we have to do in proving program correctness" [Maurer (1979), p. 627].

Two Kinds of Complexity. DeMillo, Lipton, and Perlis, however, cannot be quite so readily dismissed. Their argument is rather intriguing:

> No programmer would agree that large production systems are composed of nothing more than algorithms and small programs. Patches, *ad hoc* constructions, bandaids and tourniquets, bells and whistles, glue, spit and polish, signature code, blood-sweat-and-tears, and, of course, the proverbial kitchen sink – the colorful jargon of the practicing programmer seems to be saying something about the nature of the structures he works with; maybe theoreticians ought to be listening to him. [DeMillo, Lipton, and Perlis (1979), p. 277]

Thus, it appears to be because most real software (not contrived merely for academic purposes) tends to consist of a lot of error messages and user interfaces – "*ad hoc*, informal structures that are by definition unverifiable" – that they want to view the verificationist position as far removed from the realities of programming life [DeMillo, Lipton, and Perlis (1979), p. 277]. But the strong arguments advanced on both sides of this issue suggests that two different conceptions that reflect different dimensions of programming complexity may have become intricately intertwined.

Maurer's position, for example, tends to support a view of complexity for which more complex programs consist of less complex programs interacting according to some specific arrangement, a conception that is perfectly compatible with the "scaling-up" argument that DeMillo, Lipton, and Perlis are disparaging. This conception might be described as follows:

(D17) cumulative complexity =df the complexity of larger programs arising when they consist of (relatively straightforward) arrangements of smaller programs;

as opposed to an alternative conception that appears to be very different:

(D18) patchwork complexity =df the complexity of larger programs arising when they consist of complicated, *ad hoc*, peculiar arrangements of smaller programs.

The verificationist attitude thus appears to be appropriate to cumulative complexity, but far less adequate – perhaps hopelessly inappropriate – for cases of patchwork complexity, while the anti-verificationist position has the opposite virtues in relation to cases of both of these kinds, respectively.

The differences that distinguish large from small programs in the case of cumulative complexity appear to be differences of degree, while those that distinguish cumulative from patch-work complexity appear to be differences in kind. But that should come as no surprise. The differences between "large" and "small" programs often can be measured in terms of the numbers of lines that they contain, the number of macros or other subprograms that they employ, and the like. Moreover, differences in degree with respect to measureable magnitudes, such as height, weight, or bank balance, are not precluded from qualifying as differences in kind. To suppose that there cannot be any "real" difference between various things (such as being rich and being poor) simply because there are innumerable intermediate degrees of difference between them is known as "the slippery slope fallacy".

If this distinction is appropriately drawn, then the verification approach, in principle, would appear to apply to small programs and to large programs when they exemplify cumulative complexity, but would not apply – or only in part – to those that exhibit patchwork complexity. The "in principle" condition is important here, moreover, because DeMillo, Lipton, and Perlis offer reasons to doubt that large programs ever are or ever will be subject to verification, even when they do not qualify as "patchwork complex" programs:

> The verification of even a puny program can run into dozens of pages, and there's not a light moment or a spark of wit on any of those pages. Nobody is going to run into a friend's office with a program verification. Nobody is going to sketch a verification out on a paper napkin. Nobody is going to buttonhole a colleague into listening to a verification. Nobody is ever going to read it. One can feel one's eyes glaze over at the very thought. [DeMillo, Lipton, and Perlis (1979), p. 276]

This enchanting passage is almost enough to beguile one into the belief that a program verification is among the world's most boring and insipid objects. Yet even if that were true – even if it destroyed all prospects for a social process of program verification parallel to that found in mathematics – it would not establish that the potential for their production is therefore pointless and without value, when the purpose that might be served is clearly understood.

Indeed, one of the most important insights that can be gleaned from this debate is an appreciation of the role of formal demonstrations in both mathematical and programming contexts. For while it is perfectly appropriate for DeMillo, Lipton, and Perlis to emphasize the historical truth that the vast majority of mathematical theorems and computer programs never will be subjected to the exquisite pleasures of formal validation or of program verification, it remains enormously important, nonetheless, that as a theoretical possibility, those theorems and those programs could have been or could still be subjected to the critical scrutiny such a thorough examination would provide. Indeed, from this point of view, the theoretical possibility of subjecting them to rigorous appraisal appears to be more important than its actual exercise.

It is this potentiality, whether or not it is actually exercised, that offers an objective foundation for the intersubjective evaluation of knowledge claims: *z* knows that something is the case only when *z*'s belief that that is the case can be supported deductively, inductively, or perceptually, not therefore as something that has been done but as something that could be done were it required. This may be called "an examiner's view" of knowledge [cf. Hacking (1967), p. 319]. A fascinating illustration of this attitude has been expressedby R. C. Holt, who discusses what he takes to be "the three C's" of formal programming specification: clarity, completeness, and consistency. He suggests that the formal verification of program correctness represents an idealization "(much as certain laws of physics are idealizations of the motion of bodies of mass). This idealization encourages careful, proficient reasoning by programmers, *even if they choose not to carry out the actual mathematical steps in detail*" [Holt (1986), pp. 24-25]. This may be called an examiner's view of programming.

THEOREMS, ALGORITHMS, AND PROGRAMS

The argument that has gone before, however, depends on an assumption that DeMillo, Lipton, and Perlis seem to be unwilling to grant, namely: the presumption that *programs* are like *theorems* from a certain point of view. That some analogy exists between theorems and programs is a tempting inference, not least of all given their own analogy with mathematics. Suppose we consider the most (seemingly) obvious comparison that could be drawn:

	Mathematics:	*Programming*:
Objects of Inquiry:	Theorems	Programs
Methods of Inquiry:	Proofs	Verifications

Fig. 36. A Plausible Analogy.

Thus, from this perspective, proofs are to theorems as verifications are to programs – demonstrations of their truth or of their correctness. Although there are other aspects of various programs, including heuristics, especially, the focus here is upon the differences between programs and algorithms as objects of verification. As crude generalizations or as "rules of thumb" that admit of exceptions but are useful, nonetheless, heuristics are discovered by inductive procedures rather than by deductive means. [Cf. Nilsson (1980), Pearl (1984), and Romanycia and Pelletier (1985).]

This analogy, however, might not be as satisfactory as it intially seems. Consider, for example, that programs can be viewed as functions from inputs to outputs. In this case, the features of mathematical proofs which serve as counterparts to inputs, outputs, and programs, respectively, are premises, conclusions, and rules of inference. Thus, in place of the plausible analogy with which we began, a more adequate conception emerges:

	Mathematics:	*Programming*:
Domain:	Premises	Inputs
Function:	Rules of Inference	Programs
Range:	Theorems	Outputs

Fig. 37. A More Likely Analogy.

The acceptability of this analogy itself thus depends upon interpreting programs as well as rules of inference as *functions* (from a domain into a range). This analogy is still compatible with the first, however, so long as it is understood that rules of inference are used to derive theorems from premises, while programs are used to derive outputs from inputs. Thus, to the extent to which establishing that a certain theorem follows from certain premises (using certain rules of inference) is like establishing that a certain output is generated by a certain input (using a certain program), proving a theorem does appear to be like verifying a program.

The Deeper Disanalogy. To discern a deeper disanalogy, therefore, it is necessary to realize that, while algorithms satisfy the conditions specified above (of qualifying as functions from a domain into a range), programs need not do so. One reason emerges from considerations of complexity, especially when programs are patchwork complex programs, because programs of this kind have idiosyncratic features that make a difference to the performance of those programs, yet are not readily amenable to deductive procedures. The principal claim advanced by DeMillo, Lipton, and Perlis with respect to the issue before us, after all, is that patchwork complex programs have *ad hoc*, informal aspects – such as error messages and user interfaces – that are "by definition" unverifiable [DeMillo, Lipton, and Perlis (1979), p. 277]. In cases of this kind, presumably, no axioms that relate these special features of programs to their performance when executed are available, making verification impossible.

Another reason, which is more general, however, arises from features of other programs that are obviously intended to bring about the performance of special tasks by the machines that execute them. Illustrations of such tasks include the input and output behavior that is supposed to result as a causal consequence of their execution. Thus, the IBM PC manual for Microsoft BASIC defines various "I/O Statements" in the following fashion:

(A) *Statement*:	*Action*:
BEEP	Beeps the speaker.
CIRCLE (x,y) r	Draws a circle with center x,y and radius r.
COLOR b,p	In graphics mode, sets background color and pallette of foreground colors.
LOCATE row,col	Positions the cursor.
PLAY string	Plays music as specified by string.

In cases of this kind, presumably, although these "statement" commands are expected to produce their corresponding "action" effects, it should not be especially surprising that the prospects for their verification are problematic.

Advocates of program verification, however, might argue that commands like these are special cases. They are amenable to verification procedures, not in the sense of absolute verifiability, but rather in the sense of relative verifiability. The programs in which these commands appear could be subject to verification, provided special "causal axioms" are available relating the execution of these commands to the performance of corresponding tasks. The third reason, therefore, ought to be even more disturbing for advocates of program verification. For it should be evident that even the simplest and most commonplace program implicitly possesses exactly that same causal character. Consider, for example, the following program written in Pascal:

```
(B) program simple (output);
    begin
      writeln ('2 + 2 =', 2 + 2);
    end.
```

This program, of course, instructs the machine to write "2 + 2 =" on a line, followed by its solution, "4", on the same line. For either of these outcomes to occur, however, obviously depends upon various different causal factors, including the characteristics of the compiler, the processor, the printer, the paper, and every other component that influences this program's execution.

Programs as Causal Models. These considerations support the necessity to distinguish "programs" as encodings of algorithms from the logical structures that they represent. A program, after all, is a particular implementation of an algorithm in a form that is suitable for execution by a machine.

In this sense, a program, unlike an algorithm, qualifies as a causal model of a logical structure of which a specific algorithm may be a specific instance. The consequences of this realization, however, are enormous, insofar as causal models of logical structures need not have the same features that characterize those logical structures themselves. Algorithms, rather than programs, appear to be the appropriate candidates for analogies with *pure mathematics*, while programs bear comparison with *applied mathematics* instead (where axioms in applied mathematics, unlike those in pure mathematics, run the risk of observational and experimental disconfirmation).

From this point of view, it becomes possible to appreciate why DeMillo, Lipton, and Perlis accentuate the role of program testing in their argument:

> It seems to us that the only potential virtue of program proving [verification] lies in the hope of obtaining perfection. If one now claims that a proof of correctness can raise confidence, even though it is not perfect or that an incomplete proof can help one locate errors, then that claim must be verified! There is absolutely no objective evidence that program verification is as effective as, say, *ad hoc* theory testing in this regard. [DeMillo, Lipton, and Perlis (1979), p. 630]

Indeed, were it not for the presumption that "programs" are not "algorithms" in some fundamental respects, these remarks would be very difficult – if not altogether impossible – to understand. But when an assumption is made that:

(D19) algorithm =df a logical structure of the type function suitable for the derivation of inputs when given outputs;

(D20) program =df a causal model of an algorithm obtained by implementing that function in a form that is suitable for execution by a machine;

it is no longer difficult to understand why they should object to the conception of "program verification" as an inappropriate and unjustifiable exportation of a deductive procedure that is applicable to theorems and algorithms for the purpose of evaluating causal models that are executed by machines.

The phrase "causal model", of course, has been bestowed upon entities as diverse as *scientific theories* (such as classical mechanics and relativity theory), *physical apparatus* (such as arrangements of ropes and pulleys), and even *operational definitions* (such as that IQs are what IQ-tests test). Different disciplines tend to generate their own special senses [cf. Hesse (1966), Heise (1975), and Suppe (1977)]. The use of the term "causal model" as it is intended here, should not be misunderstood. Programs qualify as causal models because they implement algorithms in special forms that can exercise causal influence when they are loaded into suitable machines.

Abstract and Target Machines. The conception of computer programs as causal models and the difference between programs and algorithms no doubt deserve elaboration, especially insofar as there are various senses in which something might or might not qualify as a "program" or as a "causal model". Notice, first of all, that the concept of a program is highly ambiguous, since the term "program" may be used to refer to (i) algorithms, (ii) encodings of algorithms, (iii) encodings of algorithms that can be compiled, or (iv) encodings of algorithms that can be compiled and executed by a machine. And there are other senses of "program" as well [cf. Moor (1988), p. 42, for a different set of five distinctive senses]. As an effective decision procedure, moreover, an algorithm is more abstract than a program, since the same algorithm might be implemented in various specific forms suitable for execution by various specific machines by using various programming languages. From this perspective, the senses of "program" defined by (ii), (iii), and (iv) provide conceptual benefits that definition (i) does not. Indeed, were "program" defined by (i), programs could not fail to be verifiable.

The second sense, it appears, is of special importance within this context, especially in view of the distinction between "abstract machines" and "target machnines". Recall that source programs are written in high-level languages that simulate abstract machines, whose instructions can be more easily composed than can those of the target machines that ultimately execute them. It is entirely possible, in sense (ii), therefore, to envision the composition of a "program" as involving no more than the encoding of an algorithm in a programming language, no matter whether that program is now or ever will be excuted by a machine. Indeed, it might be said that composing a "program" involves no more than the encoding of an algorithm in a programming language, even if that language cannot be executed by any machine at all. An intriguing instance of this state of affairs is illustrated by the "mini-language" CORE introduced by Michael Marcotty and Henry Ledgard (1986) as a means for explaining the characteristic features of programming languages without encountering the complexities involved in discussions of Pascal, Lisp, etc. In cases of this kind, it may be said that these languages reflect features of abstract machines for which there exist no actual target-machine counterparts.

The crucial difference between "programs" in senses (i) and (ii) and "programs" in senses (iii) and (iv), therefore, is that (i) and (ii) could be satisfied merely by reference to abstract machines, while (iii) and (iv) require the existence of target machines. It might be argued that, in the case of mini-languages like CORE, for example, the abstract machine *is* the target machine. But this contention overlooks the difference at stake here, for an "abstract machine" no more qualifies as a machine than an "artificial flower" qualifies as a flower! Compilers, interpreters, processors and the rest are properly characterized as *physical systems* as entities that are in space/time and between which causal relations can obtain. Abstract machines, by comparison, are properly characterized as *abstract entities* as systems not in space/time and for which only logical relations obtain. It follows that, in senses (i) and (ii), the intended interpretations of "programs" (what the entities and relations they represent are supposed to stand for) are abstract machines that are not supposed to have physical-machine counterparts. But in senses (iii) and (iv), the intended interpretations of "programs" are abstract machines that are supposed to have physical-machine counterparts. And this difference is critical: it corresponds to that between definitions (D19) and (D20).

On the basis of these definitions, it should be evident that algorithms – and "programs" in senses (i) and (ii) – are subject to absolute verification by means of deductive procedures. This possibility arises because the properties of abstract machines that have no physical-machine counterparts can be established by definition, i.e., through stipulations or by conventions, which might be formalized by means of program rules of inference or by means of primitive program axioms. In this sense, the abstract machine under consideration simply *is* the abstract entities and relations thereby specified. But programs, by comparison – "programs" in senses (iii) and (iv) – are merely subject to relative verification, at best, by means of deductive procedures. Their differences from algorithms occur precisely because, in these cases, the properties of the abstract machines that they represent, in turn, stand for physical machines, whose properties can only be established inductively. With programs, unlike algorithms, there are no "program rules of inference" or "primitive program axioms" whose truth is ascertainable "by definition".

In either case, however, all these rules and axioms relate the occurrence of an input *I* to the occurrence of an output *O*, which can be written in the form of claims that input *I* brings about output *O* using causal conditionals:

(C) $I =u\Rightarrow O$;

which asserts that input of kind *I* invariably brings about output of kind *O*. Thus, the occurrence of output described by '*O*' may be inferred from the occurrence of input described by '*I*' together with a premise of that form in light of the following rule, which is a purely deductive rule of inference:

(D) From '$I =u\Rightarrow O$' and '*I*', infer '*O*';

since the conclusion '*O*' clearly follows from the premises '$I =u\Rightarrow O$' and '*I*'. The difference between algorithms and programs, from this point of view, is that patterns of reasoning of form (D) are absolutely verifiable in the case of algorithms but are only relatively verifiable in the case of programs – a difference that reflects the fact that assertions of form (C) themselves can be established by deductive procedures – as definitional stipulations – with respect to algorithms but can only be ascertained by inductive procedures – as causal generalizations – in the case of programs, properly understood.

THE AMBIGUITY OF "PROGRAM VERIFICATION"

When entertained from this point of view, the fundamental difficulty encountered in attempting to apply deductive methodology to the verification of programs does not appear to arise from either the idiosyncracy of various features of those programs or from the inclusion of instructions for special tasks to be performed. Both types of cases are matters that could be dealt with by introducing special rules and special axioms that correspond to the *ad hoc* features of those patchwork complex programs or to the special behavior that is supposed to be exhibited in reponse to such special commands. The specific inputs '*I*' and the specific outputs '*O*', after all, can be taken to cover these special kinds of cases. Let us therefore take this for granted in order to provide the strongest case possible for the verificationist position, thereby avoiding any chance of being charged with attacking a straw man.

As we have discovered, the crucial problem confronting program verification is establishing the truth of claims of form (C). This might be done in two possible ways. The first is to interpret rules and axioms of form (C) as definitional truths reflecting the necessary behavior of an abstract machine thereby defined. The other is to interpret rules and axioms of form (C) as empirical claims concerning the possible behavior of some target machine thereby described. Interpreted in the first way, however, the performance of an abstract machine can be conclusively verified, but it possesses no significance at all for the performance of any physical system. Interpreted in the second way, the performance of an abstract machine possesses significance for the performance of a physical system, but it can no longer be conclusively verified. And the reason by now should be obvious; for programs are subject to "relative" rather than "absolute" verification only in relation to "rules" and "axioms" in the form of causal generalizations as premises – empirical assertions whose truth can never be established with certainty!

The very idea of "program verification" thus trades on an equivocation. Interpreted in senses (i) and (ii), there is no special difficulty that arises in "verifying" that output '*O*' follows from input '*I*' as a logical consequence of axioms of the form '$I =u\Rightarrow O$'. In this case, the absolute verification of an abstract machine is logically possible and is not in principle problematic. Under such an interpretation, however, nothing follows from the verification of a "program" concerning the performance of any physical system. Interpreted in senses (iii) and (iv), by comparison, that output '*O*' follows from input '*I*' as a logical consequence of axioms of the form '$I =u\Rightarrow O$' cannot be subject to absolute verification, precisely because the truth of these axioms depends upon the causal properties of physical systems, whose presence or absence is only ascertainable by means of inductive procedures. In this case, the absolute verification of an abstract machine is logically impossible, because its intended interpretation is some target machine whose behavior might not be described by those axioms, whose truth can only be established by induction.

This conclusion strongly suggests that the conception of programming as a mathematical activity requires qualification in order to be justifiable. For while it follows from the axioms for the theory of natural numbers that, say,

(E) $2 + 2 = 4$,

the application of that assertion – which may be true of the abstract domain to which its original interpretation refers – for the new purpose of describing the behavior of physical things like alcohol and water need not remain true:

(F) 2 units of water + 2 units of alcohol = 4 units of mixture.

For while the abstract assertion (E) is true, the empirical assertion (F) is not. The difference involved here, therefore, is precisely that between *pure* and *applied* mathematics [cf. Hempel (1949a) and (1949b)]. If the function of a program is merely to satisfy the constraints imposed by an abstract machine for which there is no intended interpretation with respect to any physical machine, then the behavior of that system can be subject to absolute verification. The "four basic principles" of Hoare (1969) are then true, because:

(1′) computers are *applied* mathematical machines,
(2′) computer programs are *applied* mathematical expressions,
(3′) a programming language is an *applied* mathematical theory, and
(4′) programming is an *applied* mathematical activity;

But if the function of a program is to satisfy the constraints imposed by an abstract machine for which there *is* an intended interpretation with respect to some physical system, the behavior of that system cannot be subjected to conclusive absolute verification but requires instead empirical inductive investigation to support inconclusive relative verifications. In cases of such a kind, Hoare's four principles are false and require displacement as follows:

(1′) computers are *applied* mathematical machines,
(2′) computer programs are *applied* mathematical expressions,
(3′) a programming language is an *applied* mathematical theory, and
(4′) programming is an *applied* mathematical activity;

where assertions in applied mathematics, unlike those in pure mathematics, necessarily run the risk of observational and experimental disconfirmation.

Programs as Applied Mathematics. The differences between pure and applied mathematics are very great, indeed. As Einstein remarked, insofar as the laws of mathematics refer to reality, they are not certain, and insofar as they are certain, they do not refer to reality. DeMillo, Lipton, and Perlis likewise want to maintain that to the extent to which programming practice goes beyond the evaluation of algorithms, it cannot rely upon verification, and to the extent to which it can rely upon verification, it cannot go beyond the evaluation of algorithms. Indeed, from the perspective of the theory of knowledge, their position makes excellent sense; for the investigation of the properties of programs (when properly understood) falls within the domain of inductive methodology, while the investigation of the properties of algorithms falls within the domain of deductive methodology. As we have discovered, these are not the same. Their capabilities are altogether different.

Since the behavior of algorithms can be known with certainty (within the limitations of deductive procedures), but the behavior of programs can only be known with uncertainty (within the limitations of inductive procedures), the degree of belief (or "strength of conviction") to which specific algorithms and specifc programs may be entitled can vary greatly. A hypotheical scale once again could be constructed, this time for measuring "degrees of belief":

```
—  conviction (complete certainty)
+  1.0
+
+
+       various degrees
+  0.5        of
+        partial belief
+
+
+  0.0
—

   skepticism (complete uncertainty)
```

Fig. 38. A Measure of Subjective Belief.

Thus, to the extent to which a person can properly claim to be rational with respect to his beliefs, there should be an appropriate correspondence (which need not necessarily be an identity) between his degree of subjective belief that something is the case (as reflected by Figure 38) and the measure of objective evidence in its support (as reflected by Figure 35). Otherwise, such a person does not distribute their degrees of belief in accordance with the available evidence and is to that extent irrational [cf. Fetzer (1981), Ch. 10].

The conception of a program as a causal model suitable for execution by a machine reflects the interpretation of programs as causal factors that interact with other causal factors to bring about a specific output as an effect of the introduction of some specific input. As everyone appears willing to admit, the execution of a program qualifies as causally complex, insofar as even a correct program can produce "wildly erratic behavior . . . if only a single bit is changed". The reason that the results of executing a program cannot provide deductive support for the evaluation of a program [as Hoare (1986), p. 116, acknowledges], moreover, is that the behavior displayed by a causal system is an effect of the complete set of relevant factors whose presence or absence made a difference to its production. This situation, of course, has an analog with respect to inductive procedure, since it is a fundamental principle of inductive methodology that measures of evidential support should be based upon the complete set of relevant evidence that is currently available, where any finding whose truth or falsity increases or decreases the evidential support for a conclusion is evidentially relevant.

The principle of maximal specificity for causal systems and the principle of total evidence for inductive inference are mutually reinforcing. Another reason why reasoning about programs tends to be inductive and uncertain, after all, is that the truth of any nomological generalization about a causal system depends upon its complete specification, which can be very difficult – if not practically impossible – to ascertain. In the absence of knowledge of this kind, however, the best information available has to be conjectural. Indeed, if the knowledge that deductive warrants can provide is said to be "perfect", then our knowledge of the behavior of causal systems must always be "imperfect", experimental and tentative (like physics) rather than "perfect", demonstrative and certain (like mathematics). It is therefore ironic to realize that DeMillo, Lipton, and Perlis' position implicitly supports "the fuzzy notion that the world of programming is like the world of Newtonian physics" – not in its subject matter, needless to say, but in its methodology.

Testing Computer Programs. At least two lines of defense might be offered against this conclusion, one of which depends upon the possibility of an *ideal* programmer, the other upon the prospects for verification by *machine*. The idea of an ideal programmer is that of some person who knows as much about algorithms, programming, and computers as there is to know. When this programmer is satisfied with a program, by hypothesis, that program is "correct". The catch here is two fold. First, how could any programmer possess all the knowledge that would be required to be an "ideal programmer"? Unless this person were God, we may safely assume that the knowledge that he possesses has been ascertained by means of the usual methods, including (fallible) inductive reasoning about the future behavior of complex systems based upon evidence about their past behavior. An alternative would be to emphasize those social processes whereby more than one "very good" programmer might come to accept or to believe in the correctness of a program. But we have already discovered that this will not do. After all, the sentence, "This program reflects what programmer z believes are the right commands", can be true when the sentence, "This program reflects the right commands", happens to be false. The former is a claim about z's *beliefs*, the latter about a *program* – even if the individual z is replaced by a programming group Z.

Second, even a correct program is but one feature of a complex causal system. The performance of a computer while executing a program depends not only upon the software but also upon the firmware, the hardware, the input-output-devices, and all the rest. While it would not be mistaken to maintain that, *ceteris paribus*, if these other components perform as they should, the system will perform as it should, such claims are not testable. The emphasis here is upon the word "should". Since the outputs that result from various inputs are complex effects of an interacting arrangement of software, firmware, hardware, and so on, the determination that some specific component of such a system functions properly ("as it should") depends upon assumptions concerning the specific states of each of the others, in the absence of which, strictly speaking, no program as such can be subjected to test. We thus encounter the epistemic interdependence of software, firmware, hardware, printer, paper, and input-output ascriptions for computer behavior parallel to the epistemic interdependence of motive, belief, ethics, ability, capability, and opportunity ascriptions which obtains for human behavior.

Even when the specific states of the relevant components have been specified, moreover, the production of output O when given input I on one occasion provides no guarantee that output O would be produced by input I on some other occasion. Indeed, the problem is not only that any inference from the past to the future would have to be inductive, but also that the type of system created by the interaction of these component parts itself could be probabilistic rather than deterministic. Axioms of deterministic form (C) might have to be displaced by others of a probabilistic form:

(G) $I =n\Rightarrow O$;

which asserts that input of kind I probably brings about output of kind O. Given this causal axiom, the occurrence of output 'O' can be probabilistically derived from the occurrence of input 'I' by an inference rule of form (H):

(H) From '$I =n\Rightarrow O$' and 'I', probably 'O';

where the conclusion 'O' does not follow from but rather receives inductive support from the causal axiom '$I =n\Rightarrow O$' and input 'I'. Those premises do not guarantee the truth of that conclusion but render it "probable" instead.

A striking example of the kind of difficulty that can be encountered in cases of this type is illustrated by the distinction between "hard" and "soft" errors, which has arisen especially clearly with respect to quantum contexts:

> In the 1980s, a new generation of high-speed computers will appear with switching devices in the electronic components which are so small they are approaching the molecular microworld in size. Old computers were subject to "hard errors" – a malfunction of a part, like a circuit burning out or a broken wire, which had to be replaced before the computer could work properly. But the new computers are subject to a qualitatively different kind of malfunction called "soft errors" in which a tiny switch fails during only one operation – the next time it works fine again. Engineers cannot repair computers for this kind of malfunction because nothing is actually broken. [Pagels (1982), p. 125]

While Hoare has acknowledged the potential unreliability of the electronics, which covers cases of soft errors, his position is not therefore more defensible. For it crucially depends upon an antecedent stipulating a condition that, in principle, can never be satisfied. Consider the following passage:

> When the correctness of a program, its compiler, and the hardware of the computer have all been established with mathematical certainty, it will be possible to place great reliance on the results of the program, and predict their properties with a confidence that is limited only by the reliability of the electronics. [Hoare (1969), p. 579]

The correctness of a program, its compiler, and the hardware of the computer, after all, could *never* be established "with mathematical certainty"!

Taken altogether, these considerations suggest that even the idea of an ideal programmer cannot improve the prospects for program verification. Repeated tests of any causal system can provide only inductive evidence of its reliability – and can never guarantee its performance in the future! It should be clear by now that the difference between proving a theorem and verifying a program does not depend upon the presence or absence of any social process during their production, but rather upon the presence or absence of causal significance. A summary of their parallels looks like this:

	Proving Theorems:	*Verifying Programs*:
Syntactic Objects of Inquiry:	Yes	Yes
Social Process of Production:	Yes	No
Physical Counterparts:	No	Yes
Causal Significance:	No	Yes

Fig. 39. A Final Comparison.

The difference in social processing is merely a difference in practice, of-course, but the difference in causal significance is a difference in principle.

The suggestion can also be made (perhaps as a last-stand defense) that program verification might be performed by higher-level machines having the capacity to validate proofs that are several orders of magnitude more complex than those that can be mechanically verified today. This notion implicitly raises issues of mentality concerning whether digital machines have the capacity for semantical interpretation as well as syntactical manipulation, which I have addressed elsewhere in this book. Even assuming their powers are limited to string processing, this is an intriguing prospect, since proofs and programs alike *are* syntactical entities. The appropriate utilization of machine verifications might even be as an inductive method. The suggestion proposed by E. Cerutti and P. Davis in an analysis of the use of computers in relation to mathematical proofs, for example, might apply:

> For machine proofs, we can (a) run the program several times, (b) inspect the program, (c) invite other people to inspect the program or to write and run similar programs. In this way, if a common result is repeatedly obtained, one's degree of belief in the theorem goes up. [Cerutti and Davis (1969), pp. 903-904; see also Tymoczko (1979), Teller (1980), and Detlefsen and Luker (1980) regarding proofs by machine.]

But it would be mistaken to suppose that machine verifications of programs can guarantee a program's performance. Indeed, the very idea generates a further sequence of questions, since we immediately encounter the problem of determining the reliability of the verifying programs themselves ("How do we verify the verifiers?"), inviting an infinite regress of verifiers of verifiers.

Thomas Tymoczko (1979), moreover, suggests that proof procedures in mathematics have three distinctive features: they are convincing, they are formalizable, and they are surveyable. His sense of "surveyability", I think, can be adequately represented by "replicability". Proofs are convincing to mathematicians because they can be formalized and replicated. Insofar as the machine verification of programs has the capacity to satisfy the desiderata of formalizability and of replicability, their (successfully replicated) results ought to constitute evidence for their correctness. This, of course, does not alter the inductive character of the support thereby attained, nor does it increase the range of potential application of program verification. But it would be foolish to doubt the importance of computers in extending our reasoning capacities, just as telescopes, microscopes, and other innovations have extended our sensible capacities. Verifying programs, after all, could also be published (just as proofs are published) in order to be subjected to the criticism of the community – even by means of computer trials! That does not alter the fact that machine verifications cannot guarantee the performance that will result from executing a program as one more form of our dilemma.

Complexity and Reliability. From this perspective, the admonitions advanced by DeMillo, Lipton, and Perlis against the pursuit of perfection when perfection cannot be attained are clearly telling in this era of dependence on technology. There is little to be gained and much to be lost through fruitless efforts to guarantee the reliability of programs if that is not a rational goal. When they appraise the situation with respect to critical cases (such as air traffic control, missile systems, and the like) in which human lives are put at risk, the ominous significance of their position appears to be undeniable:

> the stakes do not affect our belief in the basic impossibility of verifying any system large enough and flexible enough to do any real-world task. No matter how high the payoff, no one will ever be able to force himself to read the incredibly long, tedious verifications of real-life systems,

> and, unless they can be read, understood, and refined, the verifications are worthless. [DeMillo, Lipton, and Perlis (1979), p. 276]

Thus, even when allowance is made for the possibility of group collaboration, the mistaken assumption that program performance can be guaranteed could easily engender an untenable conception of the situation encountered when human lives are placed in jeopardy (as in the case of "Star Wars", to take an example at random). For if one were to assume that the execution of a program could be anticipated with the mathematical precision characteristic of demonstrative domains, then one might more readily succumb to the temptation to conclude that decisions can be made with complete confidence in the (possibly unpredictable) operational performance of complex causal systems.

Complex systems like "Star Wars", moreover, are heterogeneous arrangements of complicated components, many of which themselves combine hardware and software. They depend upon sensors providing real-time streams of data and, to attain rapid, compact processing, their avionics portions may be programmed to reprogram themselves, a type of programming that does not lend itself to the construction of program verifications. Systems such as these differ in crucial respects from elementary programs to read an ASCII file, say, to count the number of characters or of words per line. An analysis of computer-system reliability that collapsed these diverse types of systems and programs into a single "catch-all" category would be indefensibly oversimplified. It ought to be emphasized, therefore, that, in addition to the difference between absolute and relative verifiability with respect to programs themselves, the reliability of inductive testing – not to mention observational techniques – declines as the complexity of the system increases. As a rule of thumb, the more complex the system, the less likely it is to perform as desired [cf. Kling (1987)]. The operational performance of these complex causal systems should never be taken for granted and cannot ever be guaranteed.

The blunder that would be involved in thinking otherwise does not result from the presence or the absence of social processes within this field that might foster the criticism of a community, but emanates from the very nature of these objects of inquiry themselves. That one or more persons of saintly disposition might sacrifice themselves to the tedium of eternal verification of tens of millions of lines of code for the benefit of the human race is therefore beside the point. The limitations that are involved here are not merely practical: *they are rooted in the very character of causal systems themselves.* From a methodological point of view, therefore, it might be said that programs are conjectures, while executions are attempted (and all too frequently successful) refutations [in the spirit of Popper (1965) and (1972)]. The more serious the consequences that would attend the making of a mistake, the greater the obligation to insure it is not made.

In maintaining that program verification cannot succeed as a generally applicable and completely reliable method for guaranteeing the performance of a program, DeMillo, Lipton, and Perlis thus arrived at the right general conclusion for the wrong specific reasons. Still, we are all in their debt for their efforts to clarify a confusion whose potential consequences – not only for the community of computer scientists, but for the human race – cannot be overstated and had best be understood. Among the most important lessons for artificial intelligence that emerges from this inquiry, I believe, is that those who produce software products have an ethical duty to insure that their performance capabilities are properly understood – not merely by professionals in the field, who are less likely to be taken in, but by the product's on-line users. The consequences of making mistakes in a complex world can be profound. The AI profession thereby incurs special moral obligations.

One of the most significant results to be derived from studies of this kind concerns the central role of logic in formalizing and understanding different schemes for representing knowledge. Although there are those, such as Minsky, who dispute the suitability of logic for this purpose, their positions tend to derive such plausibility as they possess from focusing upon an extremely limited conception of the logical domain. The field of logic is not restricted to sentential logic, predicate logic, or even enhanced versions of predicate logic, but derives its character from the study of *logical form*, where "logical form" encompasses every formal property of any semiotic domain that might possibly be theoretically relevant to its semantical and pragmatical properties. [Cf., for example, Prior (1967a) and (1967b), but especially Cargile (1984).] The specific subjects that happen to be pursued by specific students of logic at specific times no doubt reflect matters of current and continuing interest, but the field itself has enduring significance for that of artificial intelligence.

Indeed, a far more enlightened attitude has been expressed by Michael Genesereth and Nils Nilsson in their study of the logical foundations of AI. Because AI concerns itself primarily with *declarative* knowledge (knowing that something is the case), which is expressible in sentences, rather than with *procedural* knowledge (knowing how to do something), which is not, they acknowledge the centrality of the study of logic within this domain:

> we claim that AI deals mainly with the problem of representing and using *declarative* (as opposed to *procedural*) knowledge. Declarative knowledge is the kind that is expressed in sentences, and AI needs a language in which to state these sentences. Because the languages in which this knowledge usually is originally captured (natural languages such as English) are not suitable for computer representations, some other language with the appropriate properties must be used. It turns out ... that the appropriate properties include *at least* those that have been uppermost in the minds of logicians in their development of logical languages such as the predicate calculus. [Genesereth and Nilsson (1987), p. viii]

They therefore conclude that any language that would be suitable for the representation of knowledge within AI should have an expressive power at least as great as that of predicate calculus, a result which thus appears to correspond with the principal findings at which we have arrived here.

The very success of AI itself, moreover, depends upon the capacity to utilize syntactical forms to stand for semantical content. Throughout the course of our investigations, we have discovered instance after instance and context after context in which success hinges upon the introduction of syntactical distinctions corresponding to semantical distinctions, from the parsing criterion and the substitutional criterion to Tarski's theory of truth to semantic networks to production systems to scripts and frames to the validation of proofs and the verification of programs. In case upon case, the implicit message has remained that AI can succeed to whatever extent it can succeed as a function of the ability to utilize syntactical distinctions to stand for semantical distinctions for users of AI systems. So long as the marks that are manipulated by those systems are meaningful as signs for the users of those systems, it does not matter if the same marks are not meaningful signs for those systems themselves.

These considerations further reflect the importance of soundness and completeness proofs with respect to particular classes of syntactical rules, since their *soundness* guarantees that specific applications of those rules will never carry their semantical counterparts from the true to the false, while their *completeness* guarantees that every semantical counterpart from the true to the true can be captured by certain applications of those rules. Genesereth and Nilsson therefore emphasize results of these kinds:

> Anyone who attempts to develop theoretical apparatus relevant to systems that use and manipulate declaratively represented knowledge, and do so without taking into account the prior theoretical results of logicians on these topics, risks (at best) having to repeat some of the greatest work done by the brightest minds of the twentieth century and (at worst) getting it wrong! [Genesereth and Nilsson (1987), p. viii]

Thus, computational conceptions of language can function as the foundation for AI programming precisely because the computational languages that we create have exactly those properties that we design them to have in order to reflect precisely those distinctions that we desire in these symbolic systems.

The Nature of Mind. It might therefore be appropriate, by way of conclusion, to return to the three great problems in the philosophy of mind – the nature of mind, the relation of mind to body, and the existence of other minds – in an effort to review the strengths and weaknesses of The Basic Model with which we began:

	Human Beings:	*Digital Machines*:
Domain:	Stimuli	Inputs
Function:	Processes	Programs
Range:	Responses	Outputs

Fig. 40. The Basic Model.

We know The Basic Model promises solutions for each of these problems. If it is correct, then the properties, characteristics, and features that distinguish human minds are the very same properties, characteristics, and features that distinguish computer programs. The relation of mind to body, from this perspective, is simply that of software to hardware, while the existence of other minds can be solved by Turing's Test. The results we have discovered, however, do not support such positions and tend to undermine The Basic Model. Nevertheless, if Part I conveys the negative message that the prospects for machine mentality are dim, Part II and Part III bear the positive message that the prospects for artificial intelligence are bright, especially insofar as AI success does not depend upon the existence of machine mentality at all.

Insofar as The Basic Model represents an analogical argument that compares the properties of two things or kinds of things, it ought to be rejected either (i) when such an argument is taken to be conclusive, (ii) when there are more differences than similarities between those things, or (iii) when there are few but crucial differences between them. Since The Basic Model has enormous appeal even when properly construed as an inductive rather than as a deductive argument, its inadequacies, whatever else they may be, do not appear to arise because of (i). Moroever, because human beings and digital machines are both causal systems, they seem to share far more similarities than differences, suggesting that The Basic Model does not succumb to any objections because of (ii). If there is something fundamentally wrong with The Basic Model, therefore, it must be because there are few but crucial differences that distinguish human beings from digital machines as kinds of causal systems, thus making this argument a faulty analogy because of (iii).

Indeed, that the results of this inquiry were not likely to sustain the thesis that machines have minds ought to have been expected, if only attention had been directed, not to Turing's Test, but to Turing's Machine. The other great idea associated with his name, after all, is that of a *Turing Machine* as representing the class of problems that can be solved by means of machines. A machine of this kind consists of a paper tape (arbitrarily long) and a mechanism that can perform four kinds of operations: (1) it can move the tape one space (forward or backward); (2) it can mark the tape; (3) it can erase (unmark) the tape; and (4) it can halt. A Turing Machine thus possesses the mutually exclusive and jointly exhaustive *two-state* (on/off, high/low, marked/unmarked) character distinctive of *digital* machines. But while this conception may be suitable to define the capacities of digital machines, it is not remotely suggestive of the abilities distinctive of the minds of human beings.

Notice, in particular, that nothing about the conception of the four modes of operation of Turing Machines raises the prospect that a sequence of such marks ever stands for anything for those machines. The very idea ought to seem at least faintly ridiculous at this juncture, not least of all because Turing's conception does not even encompass a control mechanism to determine when the tape is moved, when marks are made, when they are erased, and when operations stop. That responsibility, of course, ultimately falls to the program as a set of instructions for such a machine. Yet even though this is enough to capture Haugeland's sense of an automated formal system, it does not reflect enough to qualify as a human mind. A Turing Machine can operate as an automated formal system and satisfy the conditions appropriate to be a causal system, but that does not mean that these strings of marks and associated operations are a sufficient condition to infuse form with content.

The Peircean conception of a semiotic system appears to be vastly more appropriate as a foundation for understanding mentality in human beings and in other thinking things than does that of an automated formal system. The most important ingredient of this approach, of course, is that of a sign as a something S that stands for something x in some respect or other for somebody z, thereby generating a triadic relationship between (a) the sign S and that for which it stands x, (b) the somebody z and the sign S, and (c) the somebody z and that for which it stands x, as Figure 41 now suggests:

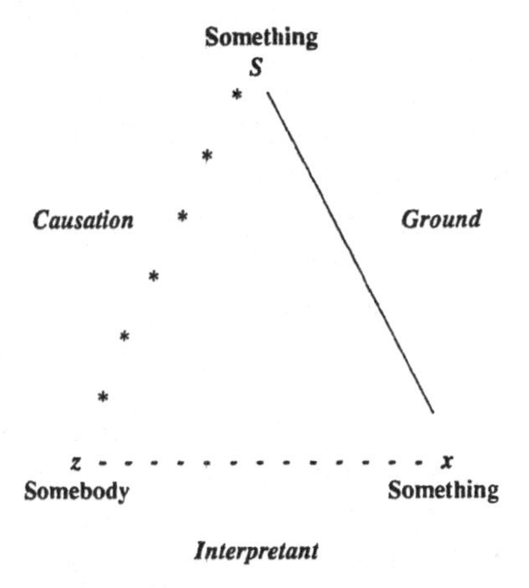

Fig. 41. The Sign Relation II.

Something S can therefore stand for something else x for somebody z only when they they are linked by corresponding relations of causation between S and z, of grounding between S and x, and of interpretant between z and x.

The relationship between a sign S and that for which it stands x reflects its "ground", which is the link by virtue of which such a something can stand for that something else. Icons, indices, and symbols thus differ from one another with respect to their grounds. The relationship between the somebody z and the something S, by comparison, is one of "causation", in the sense that the presence of that sign, under appropriate conditions, invariably or probabilistically cause the occurrence of cognition. The relationship between the somebody z and that for which that sign stands x is one of "interpretant", in the sense of z's dispositions to act in various ways under various conditions. The behavior human beings tend to display, needless to say, arises from the causal interaction of their motives, beliefs, ethics, abilities, capabilities, and opportunities as causal systems of this special kind, as explained in Chapter 3.

Signs of all three kinds can be characterized syntactically as well as semantically. The syntax of icons, for example, consists in arrangements of things of various shapes, sizes, textures, etc., in various relations. Thus, a nose of a certain shape in relation to eyes of a certain shape and color, etc., has been used to exemplify the syntax of the human face. The syntax of indices consists in arrangements of causes and of effects of various kinds, such as white smoke, bright flames, loud explosions, etc., in various relations, such as a director might create in composing special effects for his movies. And the syntax of symbols consists in arrangements of letters into words and of words into sentences in various relations, for example, in order to create poems, essays, and reports. And just as the syntax of a certain icon (such as a portrait) might resemble a certain person (such as JFK), the syntax of an index might resemble an effect (such as shouting and yelling) of a certain cause (such as military combat), and similarly for symbols.

Human Cognition. The theoretical benefits of this conception, we know, include reasonable accounts of consciousness and cognition. In particular, *consciousness* combines the ability to use signs with the capability to detect them when they are present within suitable causal proximity of a sign-user. When someone is conscious with respect to signs of certain kinds, moreover, the presence of a sign of that kind tends to bring about an instance of cognition, whose specific characteristics depends upon matters of context in the form of the other internal states of that semiotic system, including its other motives and beliefs. While "consciousness" combines ability and capability, in other words, "cognition" combines consciousness with opportunity. Thus, *cognition* occurs when a sign of a certain kind occurs within suitable causal proximity of a system which has the ability to use signs of that kind and is not inhibited from exercising that ability in detecting that sign's presence.

Yet another benefit of this pragmatical conception is that it helps us to understand what has gone wrong when different symbol users are unsuccessful in *communication*. Indeed, it should be evident that, when some speaker $z1$ utters a sentence (employs a symbol, uses a sign, etc.), successful communication with a hearer $z2$ will generally take place to the extent to which both hearer and speaker interpret that sign in a similar way. Similarity of meaning, of course, can be a matter of degree, as Chapter 3 explained. Thus, when $z1$ uses sentence S, for example, presumably he entertains S as possessing a certain content $x1$, and when $z2$ uses sentence S, presumably she entertains S as possessing a certain content $x2$, yet their interpretations might diverge:

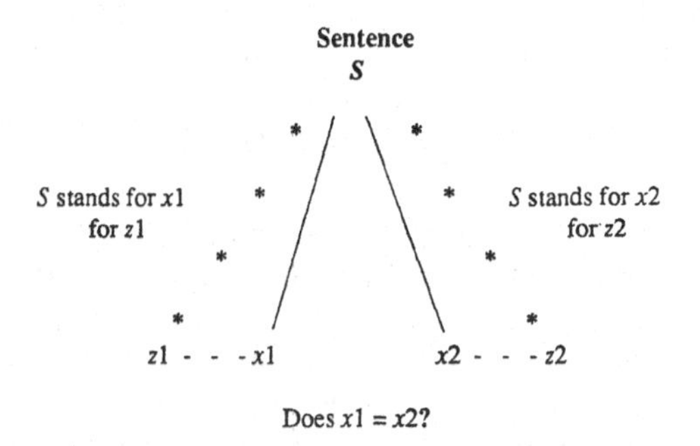

Fig. 42. Communication Situations.

Merely because S stands for $x1$ for $z1$, it does not follow that S has to stand for $x1$ for $z2$: when S stands for $x1$ for $z1$ and stands for $x2$ for $z2$, $x1$ and $x2$ may or may not be the same, a familiar dimension of human communication. [For some examples, see Shapiro and Rapaport (1987) and Rapaport (1988).]

When the same sign means more than one thing to different persons who are users of the same system of signs, then that sign is ambiguous. The proportion of a population for which the same sign may mean one thing rather than another, of course, cannot be ascertained *a priori*. Since the fundamental determination of meaning for the individual is habituation, while the fundamental determinant of meaning for the community is convention, evidently decisions must be made in the course of specification of "standard usage" and of "dictionary definitions". From the perspective of idiolects (as usages of individual sign users), of course, dictionaries are not definitive: usage is only "proper" or "deviant" in relation to an accepted standard. The principal justification for the primacy of languages (as dictionaries define them) over idiolects, therefore, is the promotion of communication within a population.

Machine Mentality. The Basic Model can be defended in various different ways from various points of view. Perhaps the most important arises from the realization that the causal sequence stimulus/process/response appears to parallel the causal sequence input/program/output. In both of these cases, a process (or program) of long duration can be linked with a stimulus (or input) of brief duration in order to bring about a response (or output) of brief duration. In both of these cases, the stimulus (or input) stands as a cause to the response (or output) in the presence of the process (or program). In both of these cases, the stimulus (or input) occurs before the response (or output) in the presence of the process (or program). And to this extent, therefore, The Basic Model would appear to be well-founded.

Indeed, the parallels are even deeper than these. The conception of consciousness for human beings, for example, combines the ability to use signs of a certain kind with the capability to exercise that ability. And the conception of cognition combines the existence of a system that is conscious in relation to signs of a certain kind with the occurrence of signs of that kind within suitable causal proximity. But digital machines can fulfill similar conditions, since a machine that can execute a Pascal program, for example, has the ability to use signs of that kind and, when it has been turned on, combines the ability to use signs of that kind with the capability to exercise that ability. And the introduction of signs of an appropriate kind within suitable causal proximity by means of their input brings about a certain output, which seems to establish a *prima facie* case in favor of machine consciousness and cognition.

In order to persist in rejecting The Basic Model as based upon an analogy that cannot be sustained, therefore, the presumption that it is well-founded, which appears to receive support from almost every direction, must be overcome. Insofar as there appear to be innumerable properties that are shared by digital machines and human beings, the only basis that remains for us to explore seems to be that there are a small number of crucial differences that distinguish them. Otherwise, the weight of the evidence suggests that, since human beings utilize (complex) causal processes from stimuli to responses, while digital machines utilize (complex) causal processes from inputs to outputs, why not concede that the parallel is exact and admit that programs are to digital machines as processes are to human beings? So if human processes are properly qualifed as "minds", why not concede the same to programs?

Thus, if the argument by analogy were explicitly unpacked, it would have to reflect an inference in two steps: first, from the parallel between stimuli and inputs and between responses and outputs to a parallel between human processes and computer programs; and, second, from the existence of human stimuli, processes, and responses as instances of human "mentality" to the existence of inputs, programs, and outputs as instances of machine "mentality":

	Human Beings:		*Digital Machines*:
Premise 1:	Stimuli	=	Inputs
Premise 2:	Responses	=	Outputs
then infer:			
Premise 3:	Processes	=	Programs
Premise 4:	(= Minds)		
and infer:			
Conclusion:			(= Minds)

Fig. 43. The Analogical Argument.

The double lines separating premises from conclusions thus indicate that these arguments are intended to be inductive (and therefore inconclusive) rather than deductive (and therefore conclusive). If something is at fault with this analogy, therefore, it must be identified and explicitly criticized.

The problem, I believe, arises with Premise 3. If the processes of human beings were the same as the programs of digital machines, the conclusion would seem to be justified – not conclusively, of course, but as inductively warranted. But if the semiotic processes of human beings involve a distinctive triadic relation between a sign (stimulus), what it stands for (response), and a sign user (process), then it can be described as follows:

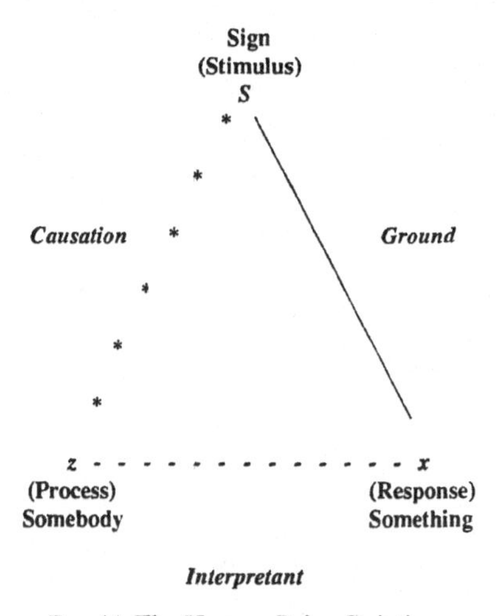

Fig. 44. The Human Being Relation.

where the relationship between "stimulus" and "response" requires a *ground* to be semiotic in kind. In other words, there may be causal connections between a cause C (call it "stimulus") and an effect E (call it "response"), but unless that causal connection obtains *because* that sign is an icon or an index or a symbol in relation to that effect (by virtue of a relation of resemblance, of cause-or-effect, or of habitual agreement), it cannot be a *semiotic* connection.

It thus becomes extremely important to understand that, even though an index stands for that for which it stands by virtue of being either a cause or an effect of that for which it stands, it is not the case that every cause has to be an index with respect to its effects. Indices are causes or effects of that for which they stand, but not all causes "stand for" their effects: something only stands for something else *for somebody* (or *for something*). Smoke, for example, stands for fire in an indexical relation for ordinary human beings, but it would be ridiculous to suppose that smoke is therefore a *cause* of fire. Red spots and an elevated temperature may stand for the measles for a physician, but it would be silly to suggest that red spots thus *cause* the measles. And while the striking of a wooden match may be a cause of its igniting, this surely does not mean that strikings *stand for* lightings for wooden matches!

In the case of digital machines, as in the case of causal systems generally, relations may obtain between causes and their effects without the existence of any relationship of *grounding* between those "causes" and those "effects":

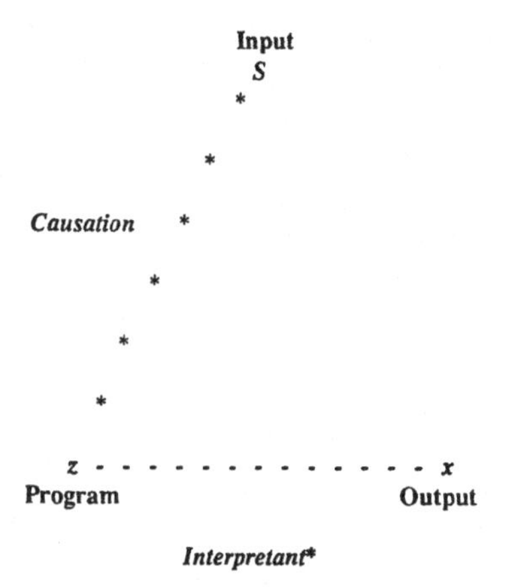

Fig. 45. The Digital Machine Relation.

Thus, if the term "interpretant" should be used at all for systems of this kind, it should be understood as lacking any semiotic significance, because of which it appears here as "interpretant*". A comparison of Figure 45 with Figure 44, I believe, should make clear the difference between causal systems that are not semiotic systems and causal systems of the special kind that qualify as semiotic systems. To the extent to which human mentality should be associated with semiotic ability, therefore, the few but crucial differences separating human beings from digital machines now seem to have been isolated and identified. The problem with the analogical argument thus turns out to be its Premise 3, without which its conclusion does not follow. And that finally establishes the fundamental, underlying inadequacy of The Basic Model itself.

MINDS AND BODIES

These findings, of course, make excellent sense from the point of view of the distinction between *simulation* and *replication*. Digital machines possess the capacity for simulating the mental processes of human beings, precisely because the causal chains linking inputs with outputs can be designed to insure that inputs that are signs for human beings yield outputs that are signs for human beings without those inputs or those outputs functioning as signs for those digital machines. In cases of this kind, Levesque's two conditions could still continue to be satisifed by the knowledge base for an AI system:

(1) there can be a one-to-one correspondence between a certain class of symbols in the KB and the class of objects of interest in the world;
(2) for every simple relationship of interest in the world, there can be a type of connection among symbols in the KB such that the relation ship holds among a group of objects in the world if and only if the appropriate connection exists among the corresponding symbols in the KB. [Levesque (1986a), p. 93]

For these conditions do not require that the kind of relationship holding between the symbols in the knowledge base and the kind of relationship that holds between the objects in the world has to be either similar or the same.

When a stimulus functions as a sign for a human being, then the response displayed by a system of that kind is brought about (at least in part) *because* that sign stands for that for which it stands, which affects its dispositions to behave in various ways under various conditions, especially those of context. When an entry functions as an input for a digital machine, however, the output exhibited by a system of that kind is *not* brought about (even in part) because that sign stands for something for that machine. For the causal linkage between inputs and outputs of digital machines are completely independent of the existence of any semiotic relationship between those inputs and those outputs by virtue of which those inputs might function as icons, as indices, or as symbols for those machines themselves rather than for their users. Thus, the types of relations which connect them are neither similar nor the same.

If the mental processes that effect the connection between stimuli (signs) and responses (behavior) in the case of human beings are neither similar to nor the same as the computer programs that effect the connection between inputs and outputs in the case of digital machines, then it may be physically impossible for digital machines to replicate the mental processes of human beings. This result, assuming that it is true, might not have been the case, if those processes were governed by different physical laws. It is not logically impossible (merely a matter of semantical convention or definitional truth) that human mental processes are not similar to or the same as digital machine programs, physical symbol systems, and automated formal systems. Once having adopted a specific conception of mentality (of minds as semiotic systems), we have discovered that digital machine programs, physical symbol systems, and automated formal systems do not appear to satisfy sufficient conditions to be minds. But our evidence cannot be conclusive.

The only way in which an argument about this matter could turn out to be "conclusive" would be if it involved deductive reasoning from premises that could not be false. In that case alone, it would be possible to derive a completely conclusive argument. So long as it is logically possible that any of our premises might be false, we are not justified in assuming that a conclusion has to be true merely because it follows from our premises. Thus, the arguments that have been provided here could lead to conclusions that are false even when they are valid, whenever the premises on which they depend might be mistaken. If digital machines are not adequately characterized as physical symbol systems or as automated formal systems, then it might still be possible for them to possess the mental abilities of human beings. But the evidence we have considered strongly suggests otherwise.

A subtle tension nevertheless remains. This chapter has claimed that that human beings and digital machines possess more similarities than differences because they are causal systems. Chapter 3 has also claimed, however, that human beings and digital machines are causal systems of "fundamentally different" kinds, which seems to imply instead that they possess more differences than similarities. My response to this tension is that it is more apparent than real. Human beings and digital machines are both *causal systems* and *of different kinds*. It is their difference as fundamentally different systems in *the strong sense* (involving their respective modes of operation) that makes the difference here. Their differences as fundamentally different systems in *the weak sense* (involving the kind of material of which they are made) are adequate to establish that machines do not possess human mentality but not that they do not have minds at all.

This point deserves elaboration. It would not be difficult to maintain that, *because* human beings are motive-belief-ethics-ability-capability-and-opportunity-types of systems, *whereas* digital machines are software-firmware-hardware-and-input/output-types of systems, *probably* the analogical argument ought to be rejected. An argument of this kind may not be trivial, but its propriety depends on whether these properties are relevant or not. While at least two criteria of relevance might be employed, they appear to offer different answers to this question. On statistical relevance criteria, it looks as though these are relevant properties, because the incidence of mentality appears to be positively correlated with their presence. On causal relevance criteria, however, it depends upon having a theory of how the properties in question are causally related. The causal approach is needed here.

In the section that follows, moreover, I want to establish that there is a more profound similarity between human beings and digital machines than their character as "fundamentally different" types of systems might suggest could possibly be the case. Indeed, this deeper parallel tends to suggest the significance of emphasizing that these are *causal systems of different kinds* rather than that these are *different kinds of causal systems*. It is therefore intended to provide the strongest possible case for denying that they have more differences than they have similarities. To establish a deep similarity, however, I must resort to formal methods, insofar as the foundation for this comparison concerns the logical form of the lawful relations that characterize systems of these types. This section, therefore, tends to impose more formal demands upon the reader than those imposed any other section of this book.

Laws of Human Beings. The behavior that results from the presence of a sign for a human being as a *semiotic system* is an effect of a complex causal system consisting of specific motives, beliefs, ethics, abilities, capabilities and opportunities. Let us refer to humans that instantiate some specific fixed set of values for the variables motives, beliefs, ethics, abilities, capabilities, especially (bearing in mind that opportunities tend to reflect whether or not the world is the way we believe it to be, i.e., as functions of the truth of beliefs), as systems of kind M (or of kinds $M1$, $M2$, ...), as appropriate. Then when stimuli and responses are of kinds S (or $S1$, $S2$, ...) and R (or $R1$, $R2$, ...), as appropriate, respectively, familiar forms of lawlike sentence must assert that subjecting any system of kind M to a stimulus of kind S would tend to bring about a response of kind R (with universal or probabilistic strength).

For those cases in which a causal connection of universal strength u obtains, there is a deterministic relationship between a stimulus of kind S and a response of kind R, which can be characterized more formally as follows:

(HL1) $(x)(t)[Mxt \Rightarrow (Sxt =u\Rightarrow Rxt^*)]$.

This asserts that, for all x and all t, if x were a system of kind M at t, then subjecting x to a stimulus of kind S at t would invariably bring about a response of kind R by x at t^* (since this is a deterministic case, by hypothesis). A logically equivalent form of (HL1), furthermore, could be fashioned by a reversal of the positions of "Mxt" and "Sxt", since the occurrence of "Rxt^*" is a complex causal consequence of their simultaneous instantiation as follows:

(HL2) $(x)(t)[Sxt \Rightarrow (Mxt =u\Rightarrow Rxt^*)]$.

This asserts that, for all x and all t, if x were subjected to a stimulus of kind S at t, then being a system of kind M at t would invariably bring about a response of kind R by x at t^*.

For those cases in which a causal connection of probabilistic strength n obtains, there is an indeterministic relationship between a stimulus of kind S and a response of kind R, which can be formally characterized as follows:

(HL3) $(x)(t)[Mxt \Rightarrow (Sxt =n\Rightarrow Rxt^*)]$.

This asserts that, for all x and all t, if x were a system of kind M at t, then subjecting x to a stimulus of kind S at t would probabilistically bring about a response of kind R by x at t^* (with a strength equal to n). And similarly for:

$$\text{(HL4)} \quad (x)(t)[Sxt \Rightarrow (Mxt =n\Rightarrow Rxt^*)],$$

which asserts that, for all x and all t, if x were subjected to a stimulus of kind S at t, then being a system of kind M at t would probabilistically bring about a response of kind R by x at t^* (with a strength equal to n).

The use of subjunctive and of causal conditionals, of course, has to be justified on logical or on ontological grounds. Conditionals of forms (HL1) and (HL3), for example, would be justified if their consequent-predicates were part of the meaning of their antecedent-predicates or if their consequent-properties were permanent properties of their antecedent-properties. Let us suppose the former were the case; then a sentence of form (HL1) would follow from a sentence of form (HL5), where (HL5) is an analytic sentence:

$$\text{(HL5)} \quad \Box(x)(t)[Mxt \rightarrow (Sxt =u\Rightarrow Rxt^*)].$$

This asserts that, necessarily, for all x and all t, if x is a system of kind M at t, then subjecting x to a stimulus of kind S at t would invariably bring about a response of kind R by x at t^*. Thus, as an analytic sentence, the truth of (HL5) follows from the adoption of a language framework **L** alone.

Whether sentences of forms such as (HL1) and (HL3) turn out to be analytic or not clearly depends upon the adoption of language frameworks of certain kinds, since whenever sentences of form (HL5) are not true within them, sentences of forms (HL1) and (HL3) will be true, when they happen to be true at all, as synthetic sentences whose consequent-properties are permanent properties of their antecedent-properties. A principle which is relevant to the adoption of such a framework, moreover, is that of *maximizing empirical content*, which suggests that analytic sentences ought to be displaced by synthetic counterparts whenever appropriate in order to promote the objective of securing hypotheses and theories of broad scope and systematic power [Fetzer (1981), pp. 290-292]. Either way, however, it is possible to explain the relationship between human bodies and their minds.

Since systems of kind M ($M1$, $M2$, ...) have been characterized in terms of their tendencies to behave in various ways under various conditions (as motives, beliefs, ethics, abilities, and capabilities), it would be appropriate to investigate the prospect of discovering some underlying neurophysiological states of kind B ($B1$, $B2$, ...), where states of kind M might turn out to be permanent properties of states of kind B. Thus, for example, the basic form,

(HL6) $(x)(t)(Bxt \Rightarrow Mxt)$,

asserts that, for all x and all t, if x were B at t, then x would be M at t; or, we might say, if x were in *brain-state* B at t, then x would be in *mind-state* M at t. Then premises in the form of sentences such as (HL6) in conjunction with others of form (HL1), (HL2), (HL3), and the like, would entail others such as:

(HL7) $(x)(t)[Bxt \Rightarrow (Sxt =u\Rightarrow Rxt^*)]$,

which asserts that, for all x and all t, if x were a system of kind B at t, then subjecting x to a stimulus of kind S at t would invariably bring about a response of kind R by x at t^*, which is a deterministic case; or others such as:

(HL8) $(x)(t)[Bxt \Rightarrow (Sxt =n\Rightarrow Rxt^*)]$,

which asserts that, for all x and all t, if x were a system of kind B at t, then subjecting x to a stimulus of kind S at t would probabilistically bring about a response of kind R by x at t^* (with a strength n), as an indeterministic case.

However, a crucial distinction has to be drawn between laws of the form (HL6), which are true when the attribute-property, M, is a permanent property of the reference-property, B, from corresponding accidental property relations that obtain when, although something, a, is in a certain brain-state B and in a certain mind-state M, M is not a permanent property of B; hence,

(HL9) Bat & Mat,

which asserts that a is in brain-state B at t and a is in mind-state M at t, yet does not assert that this relationship is nomological rather than coincidental. Nevertheless, *brain-states* as states of *brains* B^* and *mind-states* as states of *minds* M^* must not be confused, for brain states and mind states are characteristically transient rather than permanent properties of brains and minds.

Indeed, from this point of view, the key to unpacking the relationship between bodies and minds involves viewing "brains" B^* as *predispositions to acquire semiotic dispositions* M^* (M^*1, M^*2, ...) falling within some specific range (for example, English, French, ...), where which specific values within that specific range of values happen to be acquired depends upon and varies with environmental – including social and physical – factors *EF* (*EF*1, *EF*2, ...):

(HL10) $(x)(t)[B^*xt \Rightarrow (EFxt =u\Rightarrow M^*xt^*)]$.

This asserts that, for all x and all t, if x were a brain of kind B^* at t, then exposure to environmental factors of kind *EF* at t would invariably bring about the acquisition of a mind of kind M^* at t^*. The interval between t and t^*, no doubt, can involve a substantial period of time: the ability to utilize an ordinary language like English typically requires extensive linguistic training, etc.

The possession of a mind of a certain kind M^*, however, is still not enough to bring about behavior by a human being as a complex causal system, which arises from the causal influence of motives, beliefs, ethics, etc. And the acquisition of specific fixed values for variables of these kinds depends upon and varies with other environmental – including social and physical – factors EF^*:

(HL11) $(x)(t)[M^*xt \Rightarrow (EF^*xt =u\Rightarrow Mxt^*)]$.

This asserts that, for all x and all t, if x were a mind of kind M^* at t, then exposure to environmental factors of kind EF^* at t would invariably bring about the acquisition of a mental state of kind M at t^*. [For further elaboration and discussion, cf., for example, Fetzer (1988a), (1988b), and especially (1990).]

Laws of Digital Machines. Given these results for laws of human beings, let us consider the parallel prospects for laws of digital machines. The output that results from the introduction of an input for a digital machine as a *symbol system* is an effect of a complex causal system consisting of specific software, firmware, hardware, and input/output devices. Let us refer to systems that consist of some specific fixed set of values of all of these variables as systems of kind *C* (or of kind *C*1, *C*2, ...). Then when inputs and outputs are of kinds *I* (or *I*1, *I*2, ...) and *O* (or *O*1, *O*2, ...) as appropriate, respectively, corresponding forms of lawlike sentences for digital machines would assert that loading any system of kind *C* with an input of kind *I* would tend to bring about an output of kind *O* (with universal or probabilistic strength).

For those cases in which a causal connection of universal strength u

obtains, there is a deterministic relationship between an input of kind I and an output of kind O, which can be characterized more formally as follows:

(CL1) $(x)(t)[Cxt \Rightarrow (Ixt =u\Rightarrow Oxt^*)]$.

This asserts that, for all x and all t, if x were a system of kind C at t, then loading x with an input of kind I at t would invariably bring about an out put of kind O by x at t^* (because this is a deterministic case, by hypothesis). A logically equivalent form of (CL1), furthermore, could be fashioned by a reversal of the positions of "Cxt" and "Ixt", since the occurrence of "Oxt^*" is a complex causal consequence of their simultaneous instantiation as follows:

(CL2) $(x)(t)[Ixt \Rightarrow (Cxt =u\Rightarrow Oxt^*)]$.

This asserts that, for all x and all t, if x were loaded with an input of kind I at t, then being a system of kind C at t would invariably bring about an output of kind O by x at t^*.

For those cases in which a causal connection of probabilistic strength n obtains, there is an indeterministic relationship between an input of kind I and an output of kind O, which could be formally characterized as follows:

(CL3) $(x)(t)[Cxt \Rightarrow (Ixt =n\Rightarrow Oxt^*)]$.

This asserts that, for all x and all t, if x were a system of kind C at t, then loading x with an input of kind I at t would probabilistically bring about an output of kind O by x at t^* (with a strength equal to n). And similarly for:

(CL4) $(x)(t)[Ixt \Rightarrow (Cxt =n\Rightarrow Oxt^*)]$,

which asserts that, for all x and all t, if x were loaded with an input of kind$I$ at t, then being a system of kind C at t would probabilistically bring about an output of kind O by x at t^* (with a strength equal to n).

The use of subjunctive and causal conditionals, as before, has to be justified on logical or on ontological grounds. Conditionals of forms (CL1) and(CL3), for example, would be justified if their consequent-predicates were part of the meaning of their antecedent-predicates or if their consequent properties were permanent properties of their antecedent-properties. Let us suppose the former were the case; then a sentence of form (CL1) would follow from a sentence of form (CL5), where (CL5) is an analytic sentence,

$$\text{(CL5)} \quad \Box(x)(t)[Cxt \rightarrow (Ixt =u\Rightarrow Oxt^*)].$$

This asserts that, necessarily, for all x and all t, if x is a system of kind C at t, then subjecting x to an input of kind I at t would invariably bring about an outcome of kind O by x at t^*. Thus, as an analytic sentence, the truth of (CL5) follows from the adoption of a language framework **L** alone.

Since systems of kind C ($C1$, $C2$, ...) have been characterized in terms of their tendencies to behave in various ways under various conditions (of software, firmware, hardware, and input/output arrangement), it would be appropriate to consider the prospects of locating some underlying physical states of kind H ($H1$, $K2$, ...), where states of kind C might turn out to be permanent properties of states of kind H. Thus, for example, the basic form,

$$\text{(CL6)} \quad (x)(t)(Hxt \Rightarrow Cxt),$$

asserts that, for all x and all t, if x were H at t, then x would be C at t; or, in other words, if x were in *hardware-state* H at t, then x would be in *computational-state* C at t. Premises in the form of sentences such as (CL6) conjoined with others of form (CL1), (CL2), (CL3), etc., would entail others still such as:

$$\text{(CL7)} \quad (x)(t)[Hxt \Rightarrow (Ixt =u\Rightarrow Oxt^*)],$$

which asserts that, for all x and all t, if x were a system of kind H at t, then subjecting x to an input of kind I at t would invariably bring about an output of kind O by x at t^*, which is a deterministic case; or yet others such as:

$$\text{(CL8)} \quad (x)(t)[Hxt \Rightarrow (Ixt =n\Rightarrow Oxt^*)],$$

which asserts that, for all x and all t, if x were a system of kind H at t, then subjecting x to an input of kind I at t would probabilistically bring about an output of kind O by x at t^* (with the strength n), as an indeterministic case.

The crucial distinction has to be drawn again between laws of the form (CL6), which are true when the attribute-property, C, is a permanent property of the reference-property, H, from corresponding accidental property relations that obtain when, though something a is in the hardware-state H and in a certain computational-state C, C is not a permanent property of H:

(CL9) $Hat \ \& \ Cat$,

which asserts that a is in hardware-state H at t and is in computational-state C at t, without also asserting that this relationship is nomological rather than coincidental. When a mind-state and a brain-state or a computational-state and a hardware-state are simultaneous states of a corresponding system, in other words, the presence of that mind-state or of that computational-state may or may not be explained by that brain-state or by that hardware-state.

Thus, the distinction between hardware states H and computational states C parallel to that between brain states and mental states (when properly understood) requires viewing "computers" H^* as *predispositions to acquire computational dispositions* C^* (C^*1, C^*2, ...) falling within some fixed range (for example, Pascal, Lisp, ...), where which specific values within that specific range of values happen to be acquired depends upon and varies with "environmenal" – especially engineering and firmware – factors EF ($EF1$, $EF2$, ...):

(CL10) $(x)(t)[H^*xt \Rightarrow (EFxt =u\Rightarrow C^*xt^*)]$.

This asserts that, for all x and all t, if x were a computer of kind H^* at t, being subjected to environmental factors of kind EF at t would invariably bring about the acquisition of a computational disposition of kind C^* at t^*. Such a system would be programmable in some language such as Pascal, Lisp, ...

The possession of computational dispositions of a kind C^*, however, is still not enough to bring about outputs as a causal consequence of inputs, which arises from the introduction of specific instructions in the form of a computer program (for number crunching, list processing, ...). And the introduction of specific fixed values for these program variables depends upon and varies with other "environmental" – especially computer programming – factors PC:

(CL11) $(x)(t)[C^*xt \Rightarrow (PCxt =u\Rightarrow Cxt^*)]$.

This asserts that, for all x and all t, if x were of programmable kind C^* at t, being subjected to environmental conditions of kind PC at t would invariably bring about the acquisition of a specific computational state of kind C at t^*.

OTHER MINDS

Not the least of the benefits that accrue from adopting this point of view, moreover, appears to be an answer to the question of whether or not computer science properly qualifies as a science. The laws of human beings which we have considered – namely, (HL1) through (HL11) minus (HL9) – might be viewed as characterizing the domain of semiotic systems. Each of these represents an infinite class of specific instances concerning the behavior of semiotic systems of different kinds. The laws of digital machines which we have considered – namely, (CL1) through (CL11) minus (CL9) – similarly might be taken as characterizing the domain of computational systems. Each of these represents an infinite class of specific instances concerning the behavior of computational systems of different kinds. Insofar as there are laws of systems of both these kinds, there can be sciences that aim at their discovery.

The benefits that accrue from establishing laws of these kinds, no doubt, include those that other sciences enjoy, namely: the ability, under suitable conditions, to explain and to predict their instances. Consider, for example, the logical form of an explanation invoking a law of form (HL1) as follows:

$$\begin{array}{lll} \text{(HE1)} & \text{(CL):} & (x)(t)[Mxt \Rightarrow (Sxt \; {=}u{\Rightarrow} \; Rxt^*)] \\ & \text{(IC):} & Mat \; \& \; Sat \\ \hline & \text{(ES):} & Rat^* \end{array} \quad [u]$$

where an explanans consisting of a covering law (CL) and initial conditions (IC) could be invoked to explain why (or, when taken account of in time, to predict that) the phenomenon described by the explanandum sentence (ES) (will) occur(s), while the bracketed number "[*u*]" reflects the strength of the entailment relation obtaining between this explanans and its explanandum. Assuming that the conditions specified by (XXV) of Chapter 5 have been satisfied, therefore, an explanation of this kind explains why the individual named by "*a*" displayed a response of kind *R* at *t** in relation to knowledge about that individual's mental state *M* and exposure to a stimulus of kind *S* at *t*. A counterpart explanation, however, might invoke a law of form (HL7):

$$\begin{array}{lll} \text{(HE7)} & \text{(CL):} & (x)(t)[Bxt \Rightarrow (Sxt =u\Rightarrow Rxt^*)] \\ & \text{(IC):} & Bat \,\&\, Sat \\ \cline{3-3} & \text{(ES):} & Rat^* \quad [u] \end{array}$$

which could also explain why this individual *a* displayed a response of kind *R* at *t** in relation to knowledge about that individual's brain state *B* and exposure to a stimulus of kind *S* at *t* (assuming, once again, that the conditions specified by (XXV) of Chapter 5 have been satisfied). Since both apparently provide "explanations" for the same explanandum, which is to be preferred?

The answer to this question, I believe, is really not so straightforward as we perhaps assume, since it tends to depend, first, upon the kind of explanation we are after; and, second, upon the nature of our knowledge about that situation. Suppose, for example, that we want to know why Jones grimaced when Smith asked him his age. In a case of this kind, presumably, we are after something about Jones' mental state (with respect to his motives, his beliefs, his ethics, etc.) that brought about this expression, where it would be pointless to reply that Jones was in brain-state *B*! A psychological question invites a psychological answer, while a neurophysiological one does not. Thus, suppose, by comparison, that we really desire to know details about Jones' neurophysiological state. Perhaps then it might be helpful to learn that Jones has an injured eardrum, that Smith spoke a little too loudly, etc.

When behavior falls beyond the normal range (in the sense that it does-not appear to be amenable to psychological explanations that refer to normal mental states), psychological explanations that refer to abnormal mental states or neurophysiological explanations may be appropriate. We tend to interact with one another on the basis of expectations derived from our experience and reasoning based upon it. The boundaries of normality are often rather vague and indistinct: as in the case of etiquette and morality, they tend to be drawn as matters of social convention. When behavior is not explainable in terms of normal mental states, we tend to search for explanations in terms of abnormal mental states. And when behavior is not explainable in terms of abnormal mental states, then we tend to search for explanations in terms of neurophysiology to account for what's gone wrong.

There thus appears to be a correspondence here between mistakes and malfunctions; for when someone possesses the wherewithal to undertake a certain venture, yet fails to pull it off, we tend to look for mistakes which-they may have made (with respect to their beliefs, especially). So long as those with whom we deal satisfy our behavioral expectations, we tend to take for granted that their brain states are normal. We therefore appeal to mental states to account for differences in the behavior that is displayed by persons with normal brains and only appeal to neurophysiology to account for differences in behavior that cannot be explained by means of mental states. From this point of view, what we want to have explained, i.e., the pragmatic goals of explanation situations, demarcates whether an explanation of form (HE1) or an explanation of form (HE7) is appropriate.

Yet there is another reason why explanations in terms of mental states tend to take precedence over explanations in terms of brain states. When-being in mind-state *M* is a permanent property of being in brain-state *B*, the choice between kinds of explanation can be resolved on these pragmatic grounds. When being in mind-state *M* is merely a transient property of being in brain-state *B*, however, there is no choice: behavior caused by transient mental states can never be explained by reference to any brain-states to which they are not nomologically related! And the same considerations apply with reference to minds *M** as transient or as permanent properties of specific brains *B**. The only kinds of explanations *requiring* reference to brains are those invoking laws of form (HL10), which account for minds. As a consequence, the place of mentalistic explanations in daily life appears to be assured – not merely in practice, but in principle as well.

Because similar considerations obtain for explanations of the properties and of the performance of digital machines, the parallels between human-beings and digital machines in relation to laws of these various kinds are obviously very strong [as Hilary Putnam (1960) may have been the first to appreciate]. Indeed, from this point of view, The Basic Model appears to be on solid ground. If I deny that the relationship between bodies and minds is the same as that between hardware and software, therefore, it is because bodies and minds are not merely *a special kind* of hardware and software and because human beings are not just *a special kind* of automated formal system. It is therefore all the more striking to discover that, in spite of the properties that serve to differentiate between symbol systems and semiotic systems, their similarities are both extensive and substantial.

Computers and Cognition. The solutions that have been offered here, of course, are not the only ones that have been advanced to these questions about mentality and machines. It may therefore be appropriate in the final pages of this book to compare and contrast the theory elaborated here with views that others have embraced, especially with respect to the key issues. Among those whose positions deserve special consideration, from this point of view, I believe, are Terry Winograd and Fernando Flores, *Understanding Computers and Cognition* (1986); Roger Schank, *The Cognitive Computer* (1984); and Marvin Minsky, "Why People Think Computers Can't" (1982). Without suggesting that other works might not also warrant contemplation, were there world enough and time, these may serve as a representative sample of positions sufficiently similar and sufficiently different to fulfill this purpose.

Winograd and Flores (1986) represent an appropriate point of departure for, unlike Schank and Minsky, they are critical of the capacity of computers for cognition. They describe computational conceptions of cognitive systems by means of three theses, which reflect their character as "symbol systems":

(1) all cognitive systems are symbol systems, which achieve their intelligence by symbolizing external and internal situations and events and by manipulating those symbols;
(2) all cognitive processes share a basic underlying set of symbol manipulating processes; and,
(3) a theory of cognition can be couched as a program in an appropriate symbolic formalism such that the program when run in the appropriate environment will produce the observed behavior. [Winograd and Flores (1986), p. 25]

These theses appear to reflect much of the notion of symbol systems that is associated with Newell and Simon's view, which is, of course, a computational conception. And when "the observed behavior" is interpreted in terms of input and outputs that correspond to those of the systems that are modeled, they also incorporate the conception of simulation rather than of replication.

Indeed, Winograd and Flores attempt to establish that computers are incapable of replicating the mental processes of human beings. They develop a position rooted in the work of thinkers such as Heidegger, Habermas, and Gadamer, especially on the nature of language, in general, supplemented by the work of J. L. Austin (1962) and of John Searle (1969) on speech acts, in particular. Emphasizing what they take to be the essentially social and conventional character of language, they contend that human processes of language and thought have crucial features that computational procedures lack:

> Our central claim in this book is that the current theoretical discourse about computers is based on a misinterpretation of the nature of human cognition and language. Computers designed on the basis of this misconception provide only impoverished possibilities for modelling and enlarging the scope of human understanding. They are restricted to representing knowledge as the acquisition and manipulation of facts, and communication as the transferring of information. [Winograd and Flores (1986), p. 78]

In order to appreciate their position, therefore, it is important to consider why representing knowledge as "the acquistion and manipulation of facts" and communcation as "the transferring of information" is not good enough.

Winograd and Flores want to displace a conception of understanding that focuses upon the individual for one that is dominated by a social dimension. They emphasize the importance of a common background of shared assumptions involving socially-oriented patterns of behavior within human society. They stress the notions of commitment, of communication, and of humanity:

> To be human is to be the kind of thing that generates commitments, through speaking and listening. Without our ability to create and accept (or decline) commitments we are acting in a less than fully human way, and we are not fully using language. [Winograd and Flores (1986), p. 78]

Thus, much of their argument could be paraphrased in relation to machines:

> To be a computer is to be a kind of thing that cannot generate commitments through speaking and listening. It is to lack the ability to create and accept (or decline) commitments, which makes it impossible for any such thing to be fully human or to use language in fully human ways.

Their work presents many variations on these and related themes, but their fundamental objections to computational conceptions assume a similar form.

One way to bring out crucial aspects of their position might be to contrast it with those of others who emphasize the nature of language. William J. Rapaport, for example, sought to improve upon the Turing Test by proposing that the ability to understand natural language is both necessary and sufficient to possess intelligence [Rapaport (1988), p. 83]. Winograd and Flores, I believe, might readily agree; but their positions otherwise differ. Rapaport suggests an extended and detailed argument for the thesis that computational systems can understand natural language. But Winograd and Flores, by comparison, advance an extended and detailed argument for the thesis that they cannot. Rapaport maintains that, since these machines can understand natural language, they must be intelligent. But Winograd and Flores contend that, since these things cannot understand natural language, they cannot be intelligent.

The position that I have defended here takes exception to Rapaport and to Winograd and Flores with respect to the centrality of language, if only because the use of symbols reflects merely one mode whereby signs can stand for that for which they stand. The central conception of the arguments that have been developed in this work revolves about the thesis that minds are semiotic systems, of which there appear to be three fundamental types. Although I therefore reject the thesis that understanding natural language is a necessary condition for mentality (insofar as Allen's birds, Pavlov's dogs, and Skinner's pigeons are each examples of semiotic systems that possess minds without understanding natural language), I accept the view that understanding natural language is a sufficient condition for mentality. But I deny that computational systems possess the capacity to understand natural language.

It might therefore be illuminating for me to explain why, although I tend to agree with Winograd and Flores in their conclusions, I disagree with them – and with Hubert Dreyfus (1972) – about the premises of their position. It would be difficult to deny that the ordinary use of natural language is very much in the spirit of their conception: commitments are created, accepted, or declined within a social context composed of a shared background of beliefs and behavior. It would also be difficult to affirm that computers participate in commitments within a social context composed of a shared background of behavior and beliefs. To this extent, therefore, their position appears to me to be correct. But their conclusion follows only if the ordinary use of ordinary language defines the boundaries of the ability to use any language at all.

There appear to be strong and weak versions of Winograd and Flores' position. The weak version reduces to the unedifying triviality that ordinary languages are languages that people ordinarily use; since computers are not *people*, the languages that they use cannot be ordinary languages; therefore, computers cannot be intelligent. The strong version asserts the potentially significant contention that ordinary languages are inherently socially oriented in their character; since computer languages are not inherently *social*, they cannot be languages comparable to human languages; therefore, computers cannot be intelligent. What they have not shown – if I am correct, what they cannot show – is that there can be no languages that are not inherently social. For if there can be languages that are *not* inherently social – if there can be "private languages" – the languages that computers employ might yet be among them. And if they are among them, then it might yet follow that computers are intelligent when they can understand these other languages.

The Cognitive Computer. On the theory of minds as semiotic systems, the thesis that languages are inherently social does not appear to be warranted. It is not difficult to imagine individual sign-users for whom there are unique signs whose significance remains unintelligible to everyone other than themselves. Indeed, even the conception of individual language-users for whom words and sentences have distinctive meaning appears to pose no problem. The very idea of an idiolect, after all, seems to satisfy this conception as an apparent exception to Winograd and Flores' idea that all languages have to be social and conventional. Their appeal to conventions, moreover, is itself revealing, insofar as the function of conventions seems to be the promotion of communication *between* language users (which is a secondary role) as opposed to defining *what it is to be a language user* (a primary role for which they improperly employ it). The nature of language seems to be a function of habits, skills and dispositions rather than a matter of social conventions.

A completely different approach is pursued by Roger Schank (1984), who takes his objective to be the demonstration that computers can understand, so long as "understanding" itself is appropriately understood. Thus, Schank distinguishes between various different species of understanding within a spectrum, two of which he labels "making sense" and "complete empathy":

> At one end of the spectrum, we have what we shall call COMPLETE EMPATHY. This is the kind of understanding that might exist between twins, very close siblings, very old friends who know each other's every move and motivation, and other rare instances of kindred spiritedness.

> At the opposite end of the spectrum are the most basic levels of understanding, which we call MAKING SENSE. At this level of understanding, events that occur in the world are interpreted by the understander only in terms of a coherent but highly literal and restricted structure, without reference to any other understander. [Schank (1984), p. 44]

MAKING SENSE, in other words, might occur when a system can use words to describe situations properly with respect to their intensions and extensions, but COMPLETE EMPATHY would require the capacity to share someone else's experiences and attitudes (imaginatively and emotionally, at least in part).

Shank concedes that computers cannot rise to the level of COMPLETE EMPATHY, if only because they lack the essential property of humanity itself:

> What level of understanding can we hope to program a computer to achieve? Computers are not likely to have the same feelings as people, or to have the same desires and goals. They will not "be" people. The complete empathy level of understanding seems to be out of reach of the computer for the simple reason that the computer is not a person. [Schank (1984), p. 46]

Schank's position here, I think, makes a great deal of sense, not least of all because it appears to acknowledge that digital machines and human beings are fundamentally different kinds of systems: human beings, after all, are motive-belief-ethics-ability-capability-and-opportunity-kinds of systems, while digital machines are hardware-firmware-software-input/output-device-kinds of systems. To this extent, I would agree with Schank's position. But notice how Schank qualifies his commitments: he does not say that computers have *no* feelings, desires or goals. Instead, he says that they are not likely to have the *same* feelings, the *same* desires, or the *same* goals – as though they had *any* feelings, desires, or goals at all! These are curious qualifications, especially when Schank suggests that AI can survive, nonetheless:

> there is plenty of work to be done just getting computers to MAKE SENSE of the same kinds of things of which humans can make sense. Just getting a computer to MAKE SENSE of the sentence, *Bill cried when Mary said she loved John* is very difficult. It is easy to assert that computers never will be programmed to become emotional or sentimental when reading such a sentence (and there may be good reasons

> for not wanting computers to become emotional or sentimental). They certain can be programmed to MAKE SENSE of such sentences, but they won't share the human experiences of being in love or of being sad. [Schank (1984), p. 46]

There is something about these remarks – and many others that he makes – that hints, however gently, that emotionality and sentimentality are perhaps not really completely out of the picture of computer possibilities, after all. I shall refrain from criticizing this aspect of his views, however, and spare myself the burden of attacking a straw man. Indeed, his positive remarks as to what computers can surely do – such as MAKE SENSE – are more to the point.

The beauty of Schank's position, however, emanates from his analysis of yet another kind of understanding, which he refers to as COGNITIVE UNDERSTANDING. Thus, he suggests that this species lies between the other two:

MAKING SENSE ________ COGNITIVE UNDERSTANDING ________ COMPLETE EMPATHY

where a machine that operates at the level of cognitive understanding may: (1) learn or change as a result of its experiences, (2) relate its present experiences intelligently, (3) formulate new information for itself by coming to its own conclusions on the basis of experience, and (4) explain why it makes the connections that it makes "and what *thought process* was involved in making a conclusion" [Schank (1984), p. 48]. Scripts, of course, figure prominently in Schank's examples of COGNITIVE UNDERSTANDING: SAM, after all, is a suitable illustration of MAKING SENSE, and PAM tends to exemplify COGNITIVE UNDERSTANDING. Given the programming requirements of systems of these kinds, however, it is difficult to know how seriously to take Schank's words.

The crucial question, of course, is whether Schank is seriously ascribing these properties in their ordinary (literal) senses or in some extraordinary (figurative) senses. As modes of simulation, there would appear to be little to object to about Schank's descriptions, but then his words have to be interpreted figuratively rather than literally. These digital machines are surely incapable of having "experiences", even when they run Schank's programs. They can "come to their own conclusions" only if they are capable of reasoning. And they can explain their own "*thought processes*" – Schank's italics – only if they are capable of thought processes of their own. On each of these crucial issues, Schank's position simply begs the question. He displays no consideration at all for the possibility that the signs that are used by a system might be meaningful for the users of that system but not for that system itself. He simply takes it for granted that computers *can* understand!

From this point of view, I believe, the most fascinating passage in this book occurs when Schank critiques (what he takes to be) the exaggerated claims made on behalf of ELIZA, an interactive Rogerian psychotherapy program. ELIZA asks questions and responds to answers by asking further questions in an effort to diagnose patient problems. But Schank assails this program as a fraud, which appears to yield *understanding* when it doesn't:

> The point here is that the program anticipates certain key words and responds to them with certain canned responses. A computer programmed to convert the statement *I hate you* to *Why do you hate me?* doesn't understand anything, doesn't have any expectations or knowledge structures for what it says, and, most importantly, doesn't tell us anything about the nature of human understanding. [Schank (1984), p. 138. Notice, however, that such a program simulates an ability of semiotic systems of Type IV!]

Schank goes on to berate the inadequacies of this program, alleging that his own programs are qualitatively superior by concentrating "on crucial differences in meaning, not on issues of grammatical structure". But whether this could possibly be done employing the exclusively syntactical operations of a completely computational device is an issue that is never addressed. Schank not only fails to have the answers; he doesn't even understand the questions.

WHY PEOPLE THINK COMPUTERS CAN'T

There are various ways in which Schank's position might be salvaged, especially by feigning figurative rather than literal forms of speech. It could be maintained, for example, that REALLY INTELLIGENT programs always rely upon heuristics, which makes them "independent thinkers" in at least some extended sense. Perhaps the problem is the medium rather than the message. Intelligent things can be said about machines without obfuscation in the use of language. [Some of them have even come from Schank (1983).] And there are excellent reasons for pursuing AI, even apart from questions of simulation and of replication. As Patrick Winston (1984) has remarked,

> Note that wanting to make computers *be* intelligent is not the same as wanting to make computers *simulate* intelligence. Artificial intelligence excites people who want to uncover principles that all intelligent information processors must exploit, not just those made of wet neural tissue instead of dry electronics. Consequently, there is neither an obsession with mimicking human intelligence nor a prejudice against using methods that seem involved in human intelligence. Instead, there is a new point of view that brings along a new methodology and leads to new theories. [Winston (1984), p. 2]

Indeed, this perspective makes excellent sense, especially with respect to the goal of designing automated devices that might perform their intended functions quite successfully and very dependably without raising any questions as to whether or not they are *really intelligent*, even if the problems that they help to solve are ones that required human intellects in the past.

Nevertheless, if Winograd and Flores (1984) advocate too strong a conception of the nature of human language, while Schank (1984) adopts too strong a conception of the character of computer capacities, Minsky (1982) comes very close to placing emphasis precisely where it properly belongs, namely: on the extent to which machines can be programmed to deal with different forms of knowledge. In his estimation, the problems of creativity, originality, and spontaneity are merely degrees of difference in magnitude of difficulty removed from the problems that common sense must resolve:

> As I see it, any ordinary person who can understand an ordinary conversation must have already in his head most of the mental powers that our greatest thinkers have. In other words, I claim that "ordinary, common sense" already includes the things it takes – when better balanced and more fiercely motivated – to make a genius. [Minsky (1982), p. 5]

Thus, Minsky suggests, those who are more successful tend to be better at *learning* and better at *managing what they learn* than most other persons.

He inventories reasons that are frequently advanced in support of the thesis that computers cannot think. He contends that computers can solve problems that have not been solved before, that computers can be as creative as human beings (because there is no substantial difference between ordinary thought and creative thought), and that computers can even simulate randomness and error in the problems they attack. He suggests that there are differences between thinking and its justification, which makes logic less appropriate for solving problems and more appropriate for displaying solutions once discovered. And in all of this there is a good deal to agree with. The problem of innovation, for example, has a variety of distinct dimensions, for creativity may merely involve doing things that no one else has done before. What makes an innovation an important or an influential one depends upon historical and contextual conditions over which machines have no control [cf. especially Fetzer (1988a), pp. 84-85].

Minsky's principal assertions, however, revolve about the problem of common sense, which entails concern for ordinary language. For it seems perfectly evident that, without the ability to understand natural language, no machine could comprehend common sense. His remarks on this score, therefore, are quite intriguing. He hesitates to pursue the nature of meaning ("I feel no obligation to define such words as 'mean' and 'understand', just because others tried it for five thousand years!"), yet astutely observes that meanings are dependent upon matters of context ("what something means to me depends to some extent on everything else I know – and no one else knows just those things in just those ways"). And he observes that the ordinary beliefs that we accept are typically inconsistent (when taken collectively) and often false (when considered individually).

More important than any of these remarks, however, is Minsky's critique of the role of formalization and the place of definition within human knowledge. He does not object to the possibility of circular definitions or even to infinite regresses, in which the meaning of language at one level is explained by language at another level. But he does suggest that there may be other than linguistic ways to understanding what language means. In discussing the meaning of numbers, for example, he indicates at least three different ways in which the number "three" might be understood: by techniques like pointing and touching, by matching members with a standard set, and by way of perceptual groupings [Minsky (1982), p. 9]. The importance of these reflections, I believe, is the emphasis they place upon the role of habits, skills, and dispositions in understanding language.

Minsky considers other issues (such as whether a machine could have "a sense of self"), the most important of which is the question of machine mentality. He also believes that these issues are related and that doubts over machine identity contribute to skepticism about machine mentality:

> I think that's partly why most people still reject computational theories of thinking, although they have no other worthy candidates. And that leads to denying minds to machines. For me, this has a special irony, because it was only after trying to understand what computers – that is, complicated mechanisms – *could* do, that I began to have some glimpses of how a *mind* itself might work. [Minsky (1982), p. 12]

The underlying difficulty with Minsky's position, I believe, begins to display itself at this juncture. While he avows his commitments to the computational conception of language and mentality, his own emphasis upon the importance of context and the role of habits, skills, and dispositions as factors affecting meaning tends to undermine the plausibility of the symbol system model. For unless context, habits, and skills are amenable to formalization, Minsky's computational conclusion appears to be inconsistent with his own non-computational evidence. They cannot both be true!

His sensitivity to the influence of background factors seems to be significant, since it suggests that every human being may have a different point of view. The theory of cognition that attends the dispositional conception, moreover, provides a perspective that can explain why it should be "a basic fact of mind: [that] what something means to me depends to some extent on everything else I know" [Minsky (1982), p. 8]. For the infinite variability in context created by unlimited combinations of motives, beliefs, ethics, abilities, capabilities, and opportunites that we can encounter implies that each of us will tend to possess a distinctive point of view as a predictable effect of genetic variability, environmental variation, and probabilistic causation. And this, in turn, suggests that the use of digital machines to reflect the perspective of human beings can never surmount the limitations inherent in adapting finite programs to infinite processes.

To the extent to which understanding common sense requires understanding natural language, Minsky's position tends to reinforce familiar difficulties that accompany prevaling computational conceptions. Moreover, he has inadvertently emphasized its greatest source of strength – the apparent absence of viable alternatives! Like many others, Minsky seems prepared to endure the storm of embracing an inadequate theory, even in the face of incompatible evidence. But this is an appropriate and scientific attitude only so long as there *are* no reasonable alternatives. The introduction of semiotic systems in place of symbol systems as models of mentality and the elaboration of dispositional conceptions in lieu of syntactical conceptions of the nature of language establishes new and significant theoretical alternatives, which promise to shed light on these old but very important problems. They should be judged by their respective capacities to explain and predict the diverse phenomena that distinguish the behavior of human beings and of digital machines. The evidence that has been offered here strongly suggests that human beings are semiotic systems but that digital machines are not. Ignoring its significance would be a great mistake.

The objective of this work has been to offer evidence for believing that The Basic Model is wrong. I have argued that human beings and digital machines are causal systems but of very different kinds; that a distinction has to be drawn between symbol systems and semiotic systems; that digital machines are symbol systems, while human beings are semiotic; that only semiotic systems are possessors of minds; and that the prevalent conceptions of AI are based upon mistakes. If the arguments that I have offered are right, The Basic Model is not merely wrong in its details but also wrong in its general conception. The Turing Test does not resolve the problem of the existence of other minds. The relation of software to hardware does not provide a solution to the mind/body problem. Mental processes are not computer programs and human beings are not digital machines. Even the conception of "programs" as functions from inputs to outputs cannot withstand scrutiny.

Nevertheless, it would be incorrect to suggest that this work has shown that computer systems could not possibly instantiate the mental processes of human beings. Indeed, unless the question is being begged, it must not be *logically impossible* for digital machines to possess the minds of human beings. What we have discovered, however, is that does not appear to be *physically possible* for digital machines to possess the minds of human beings. Were it logically impossible for digital machines to possess the minds of human beings, the question itself could be answered *a priori*. The argument elaborated here has been that minds are semiotic systems and that human beings have them but digital machines do not. The possibility that other systems – connectionist, perhaps – might have minds is not ruled out.

The net result, however, is not therefore negative, for AI can proceed in spite of these findings by embracing a more modest and rationally warranted conception of its boundaries. The representation of knowledge, the development of expert systems, the simulation and modeling of natural and behavioral phenomena, the rapid processing of vast quantities of data, information, and knowledge are destined to assume a major role in shaping the future of our species. And just as the Agricultural Revolution relieved us of a great deal of hunting and gathering and the Industrial Revolution relieved us of a great deal of physical labor, the Computer Revolution will relieve us of a great deal of mental effort. A more adequate conception of the scope and limits of AI confirms that, within the confines of what is possible, AI can provide methods and tools of immense benefit to all mankind.

REFERENCES

Ackermann, R. (1965), *Theories of Knowledge* (New York, NY: McGraw-Hill Book Company, 1965).

Allen, J. F. (1983), "Maintaining Knowledge about Temporal Intervals", *Communications of the ACM* 26 (1983), pp. 832-843.

Allen, J. F. and C. R. Perrault (1980), "Analyzing Intentions in Utterance", *Artificial Intelligence* 15 (1980), pp. 143-178.

Almeida, M. J. (1987), "Reasoning about the Temporal Nature of Narratives", *Technical Reports 87-10* (Buffalo, NY: SUNY Buffalo Department of Computer Science).

Austin, J. L. (1950), "Truth", *Proceedings of the Aristotelian Society*, Supplementary Volume 245 (1950), pp. 111-138.

Austin, J. L. (1962), *How to do Things with Words* (Oxford, UK: Oxford University Press, 1962).

Benacerraf, P. and H. Putnam, eds. (1964), *Readings in the Philosophy of Mathematics* (Englewood Cliffs, NJ: Prentice-Hall, 1964).

Black, M. (1967), "Induction", in P. Edwards, ed., *The Encyclopedia of Philosophy*, Vol. 4 (New York, NY: Macmillan Publishing Company, 1967), pp. 169-181.

Block, N., ed. (1980), *Readings in the Philosophy of Psychology*, Vol. I (Cambridge, MA: Harvard University Press, 1980).

Blumberg, A. (1967), "Logic, Modern", in P. Edwards, ed., *The Encyclopedia of Philosophy*, Vol. 5 (New York, NY: Macmillan Publishing Company, 1967), pp. 12-34.

Bochner, S. (1966), *The Role of Mathematics in the Rise of Science* (Princeton, NJ: Princeton University Press, 1966).

Boden, M. (1988), *Computer Models of Mind* (Cambridge, UK: Cambridge University Press, 1988).

Brachman, R. (1979), "On the Epistemological Status of Semantic Networks", in N. Findler, ed., *Associative Networks: Representation and Use of Knowledge by Computers* (New York, NY: Academic Press, 1979), pp. 3-50.

Brachman, R. (1983), "What IS-A Is and Isn't: An Analysis of Taxonomic Links in Semantic Networks", *COMPUTER* (October 1983), pp. 30-36.

Brachman, R. and H. Levesque, eds. (1985), *Readings in Knowledge Representation* (Los Altos, CA: Morgan Kaufmann Publishers, 1985).

Brachman, R. and B. Smith (1980), "Special Issue on Knowledge Representation", *SIGART Newsletter* Number 70.

Braithwaite, R. B. (1953), *Scientific Explanation* (Cambride, UK: Cambridge University Press, 1953).

Brown, F. M., ed. (1987), *The Frame Problem in AI: Proceedings of the 1987 Workshop* (Los Altos, CA: Morgan Kaufmann).

Buchanan, B. (1985), "Expert Systems", *Journal of Automated Reasoning* 1 (1985), pp. 28-35.

Buchanan, B. and E. Shortliffe, eds. (1984), *Rule-Based Expert Systems* (Reading, MA: Addison-Wesley Publishing Company, 1984).

Campbell, N. R. (1920), *Physics: The Elements* (Cambridge, UK: Cambridge University Press, 1920). [Republished with the title, *Foundations of Science* (New York, NY: Dover Publications, 1957).]

Cargile, J. (1984), "Classical Logic: Traditional and Modern", in J. Fetzer, ed., *Principles of Philosophical Reasoning* (Totowa, NJ: Rowman and Allanheld, 1984), pp. 44-70.
Carnap, R. (1936-37), "Testability and Meaning", *Philosophy of Science* 3 (1936), pp. 419-471: and *Philosophy of Science* 4 (1937), pp. 1-40.
Carnap, R. (1939), *Foundations of Logic and Mathematics* (Chicago, IL: University of Chicago Press, 1939).
Castaneda, H-N. (1967), "Indicators and Quasi-Indicators", *American Philosophical Quarterly* 4 (1967), pp. 85-100.
Castaneda, H-N. (1970), "On the Philosophical Foundations of the Theory of Communication: Reference", *Midwest Studies in Philosophy* 2 (1970), pp. 165-186.
Castaneda, H-N. (1972), "Thinking and the Structure of the World", *Philosophia* 4 (1972), pp. 3-40.
Cercone, N. and G. McCalla, eds. (1987), *The Knowledge Frontier: Essays in the Representation of Knowledge* (New York, NY: Springer-Verlag).
Cerutti, E. and P. Davis (1969), "Formac Meets Papus", *American Mathematical Monthly* 76 (1969), pp. 895-904.
Charniak, E. and D. McDermott (1985), *Introduction to Artificial Intelligence* (Reading, MA: Addison-Wesley, 1985).
Chomsky, N. (1965), *Aspects of the Theory of Syntax* (Cambridge, MA: MIT Press, 1965).
Chomsky, N. (1966), *Cartesian Linguistics* (New York, NY: Harper & Row, 1966).
Chomsky, N. (1986), *Knowledge of Language: Its Nature, Origin, and Use* (New York, NY: Praeger Publishers, 1986).
Church, A. (1959a), "Logic, Formal", in D. Runes, ed., *Dictionary of Philosophy* (Ames, IO: Littlefield, Adams & Company, 1959), pp. 170-181.
Church, A. (1959b), "Logistic System", in D. Runes, ed., *Dictionary of Philosophy* (Ames, IO: Littlefield, Adams & Company, 1959), pp. 182-183.
Cohen, D. (1986), *Introduction to Computer Theory* (New York, NY: John Wiley & Sons, 1986).
Cohen, I. B. (1985), *The Birth of a New Physics*, rev. ed. (New York, NY: W. W. Norton and Company, 1985).
Cohen, P. and C. R. Perrault (1979), "Elements of a Plan-Based Theory of Speech Acts", *Cognitive Science* 3 (1979), pp. 177-212.
Cullingford, R. (1981), "SAM", in R. Schank and C. Riesbeck, eds., *Inside Computer Understanding* (Hillsdale, NJ: Lawrence Erlbaum Associates, 1981), pp. 75-119.
Dancy, J. (1985), *Introduction to Contemporary Epistemology* (Oxford, UK: Basil Blackwell Publishers, 1985).
Davidson, D. (1967), "Truth and Meaning", *Synthese* 17 (1967), pp. 304-323.
DeMillo, R., R. Lipton and A. Perlis (1979), "Social Processes and Proofs of Theorems and Programs", *Communications of the ACM* 22 (1979), pp. 271-280.
Dennett, D. (1971), "Intentional Systems", *Journal of Philosophy* 68 (1971), pp. 87-106; reprinted in J. Haugeland, ed., *Mind Design* (Cambridge, MA: The MIT Press, 1981), pp. 220-242.
Detlefsen, M. and M. Luker (1980), "The Four-Color Theorem and Mathematical Proof", *Journal of Philosophy* 77 (1980), pp. 803-820.
Dijkstra, E. W. (1976), *A Discipline of Programming* (Englewood Cliffs, NJ: Prentice-Hall, 1976).

Doyle, J. (1979), "A Truth Maintenance System", *Artificial Intelligence* 12 (1979), pp. 231-272.

Dretske, F. (1985), "Machines and the Mental", *Proceedings and Addresses of the American Philosophical Association* (September 1985), pp. 23-33.

Dreyfus, H. (1972), *What Computers Can't Do* (New York, NY: Harper & Row, 1972).

Eells, E (1982), *Rational Decision and Causality* (Cambridge, UK: Cambridge University Press, 1982).

Fetzer, J. H. (1978), "On Mellor on Dispositions", *Philosophia* 7 (1978), pp. 651-660.

Fetzer, J. H. (1981), *Scientific Knowledge* (Dordrecht, Holland: D. Reidel, 1981).

Fetzer, J. H. (1983a), "Probabilistic Explanations", in P. Asquith and T. Nickles, eds., *PSA 1982*, Vol. 2 (East Lansing, MI: Philosophy of Science Association, 1983), pp. 194-207.

Fetzer, J. H. (1983b), "Transcendent Laws and Empirical Procedures", in N. Rescher, ed., *The Limits of Lawfulness* (Lanham, MD: University Press of America, 1983), pp. 25-32.

Fetzer, J. H. (1984), "Philosophical Reasoning", in J. Fetzer, ed., *Principles of Philosophical Reasoning* (Totoaw, NJ: Rowman and Allanheld, 1984), pp. 3-21.

Fetzer, J. H. (1985a), "Reduction Sentence 'Meaning Postulates'", in N. Rescher, ed., *The Heritage of Logical Positivism* (Lanham, MD: University Press of America, 1985), pp. 55-65.

Fetzer, J. H. (1985b), "Science and Sociobiology", in J. Fetzer, ed., *Sociobiology and Epistemology* (Dordrecht, Holland: D. Reidel, 1985), pp. 217-246.

Fetzer, J. H. (1986), "Methodological Individualism: Singular Causal Systems and Their Population Manifestations", *Synthese* 68 (1986), pp. 99-128.

Fetzer, J. H. (1987), "Critical Notice: Wesley Salmon's *Scientific Explanation and the Causal Structure of the World*", *Philosophy of Science* 54 (1987), pp. 597-610.

Fetzer, J. H. (1988a), "Mentality and Creativity", *Journal of Social and Biological Structures* 11 (1988), pp. 82-85.

Fetzer, J. H. (1988b), "Signs and Minds: An Introduction to the Theory of Semiotic Systems", in J. Fetzer, ed., *Aspects of Artificial Intelligence* (Dordrecht, The Netherlands: Kluwer Academic Publishers, 1988), pp. 133-161.

Fetzer, J. H. (1988c), "Program Verification: The Very Idea", *Communications of the ACM* 31 (1988), pp. 1048-1063.

Fetzer, J. H. (1989), "Language and Mentality: Computational, Representatational, and Dispositional Conceptions", *Behaviorism* 17 (1989), pp. 21-39.

Fetzer, J. H. (1990), *Philosophy and Cognitive Science* (New York: Paragon).

Fetzer, J. H. and D. Nute (1979), "Syntax, Semantics, and Ontology: A Probabilistic Causal Calculus", *Synthese* 40 (1979), pp. 453-495.

Fetzer, J. H. and D. Nute (1980), "A Probabilistic Causal Calculus: Conflicting Conceptions", *Synthese* 44 (1980), pp. 241-246.

Fodor, J. (1975), *The Language of Thought* (Cambridge, MA: MIT Press, 1975).

Fodor, J. (1980), "Methodological Solipsism Considered as a Research Strategy in Cognitive Psychology", reprinted in J. Haugeland, ed., *Mind Design* (Cambridge, MA: MIT Press, 1981), pp. 307-338.

Fodor, J. (1987), *Psychosemantics* (Cambridge, MA: MIT Press, 1987).

Freeman, D. (1983), *Margaret Mead and Samoa* (Cambridge, MA: Harvard University Press, 1983).

Garvin, P. (1985), "The Current State of Language Data Processing", *Advances in Computers* 24 (1985), pp. 217-275.
Genesereth, M. and N. Nilsson (1987), *Logical Foundations of Artificial Intelligence* (Los Altos, CA: Morgan Kaufmann, 1987).
Gettier, E. (1963), "Is Justified True Belief Knowledge?", *Analysis* 23 (1963), pp. 121-123.
Glazer, D. (1979), Letter to the Editor, *Communications of the ACM* 22 (1979), p. 621.
Goodman, N. (1965), *Fact, Fiction and Forecast* (Indianapolis, IN: Bobbs-Merrill, 1965).
Grishman, R. (1986), *Computational Linguistics* (Cambridge, UK: Cambridge University Press, 1986).
Gustason, W. and D. Ulrich (1973), *Elementary Symbolic Logic* (New York, NY: Holt, Rinehart and Winston, 1973).
Hacking, I. (1967), "Slightly More Realistic Personal Probabilities", *Philosophy of Science* 34 (1967), pp. 311-325.
Halpern, J. Y., ed. (1986), *Theoretical Aspects of Reasoning About Knowledge* (Los Altos, CA: Morgan Kaufmann).
Harris, H. and R. Severens, eds. (1970), *Analyticity* (Chicago, IL: Quadrangle Books, 1970).
Hartshorne, P. and P. Weiss (1960), *The Collected Papers of Charles S. Peirce,* Vols. 1 and 2 (Cambridge, MA: Harvard University Press).
Haugeland, J. (1981), "Semantic Engines: An Introduction to Mind Design", in J. Haugeland, ed., *Mind Design* (Cambridge, MA: MIT Press, 1981), pp. 1-34.
Haugeland, J. (1985), *Artificial Intelligence: The Very Idea* (Cambridge, MA: MIT Press, 1985).
Hayes, P. (1973), "The Frame Problem and Related Problems in Artificial Intelligence", in A. Elithorn and D. Jones, eds., *Artificial and Human Thinking* (San Francisco, CA: Jossey-Bass, 1973), pp. 45-49.
Hayes, P. (1978), "The Naive Physics Manifesto", in D. Michie, ed., *Expert Systems in the Microelectronic Age* (Edinburgh, Scotland: Edinburgh University Press, 1978), pp. 242-270.
Hayes, P. (1979), "The Logic of Frames", in D. Metzing, ed., *Frame Conceptions and Text Understanding* (Berlin, FRG: Walter de Guyer, 1979), pp. 46-61.
Hayes, P. (1985), "The Second Naive Physics Manifesto", in J. R. Hobbs and R. Moore, eds., *Formal Theories of the Commonsense World* (Norwood, NJ: Ablex Publishing Corporation, 1985), pp. 1-36.
Hayes-Roth, F., D. A. Waterman and D. Lenat, eds. (1983), *Building Expert Systems* (Reading, MA: Addison-Wesley Publishing Company, 1983).
Heise, D. R. (1975), *Causal Analysis* (New York, NY: John Wiley & Sons, 1975).
Hesse, M. (1966), *Models and Analogies in Science* (Notre Dame, IN: University of Notre Dame Press, 1966).
Hempel, C. G. (1949a), "On the Nature of Mathematical Truth", in H. Feigl and W. Sellars, eds., *Readings in Philosophical Analysis* (New York: Appleton Century-Crofts, Inc., 1949), pp. 222-237.
Hempel, C. G. (1949b), "Geometry and Empirical Science", in H. Feigl and W. Sellars, eds., *Readings in Philosophical Analysis* (New York: Appleton-Century-Crofts, Inc., 1949), pp. 238-249.
Hempel, C. G. (1952), *Fundamentals of Concept Formation in Empirical Science* (Chicago, IL: University of Chicago Press, 1952).

Hempel, C. G. (1962), "Rational Action", reprinted in N. Care and C.Landesman, eds., *Readings in the Theory of Action* (Bloomington, IN: Indiana University Press, 1968), pp. 281-305.

Hempel, C. G. (1965), *Aspects of Scientific Explanation* (New York, NY: The Free Press, 1965).

Hempel, C. G. (1966), *Philosophy of Natural Science* (Englewood Cliffs, NJ: Prentice-Hall, 1966).

Hempel, C. G. (1968), "Maximal Specificity and Lawlikeness in Probabilistic Explanation", *Philosophy of Science* 35 (1968), pp. 116-133.

Hesse, M. (1966), *Models and Analogies in Science* (Notre Dame, IN: University of Notre Dame Press, 1966).

Hoare, C. A. R. (1969), "An Axiomatic Basis for Computer Programming", *Communications of the ACM* 12 (1969), pp. 576-580 and p. 584.

Hoare, C. A. R. (1986), "Mathematics of Programming", *BYTE* (August 1986), pp. 115-124 and pp. 148-150.

Holt, R. (1986), "Design Goals for the Turing Programming Language", *Technical Report CSRI-187* (August 1986), Computer Systems Research Institute, University of Toronto.

Hutchins, W. J. (1986), *Machine Translation: Past, Present, Future* (London: Ellis Horwood Ltd.)

Jaeger, W. (1960), *Aristotle: Fundamentals of the History of His Development* (Oxford, UK: Oxford University Press, 1960).

Klemke, E., ed. (1970), *Essays on Bertrand Russell* (Urbaba, IL: University of Illinois Press, 1970).

Kling, R. (1987), "Defining the Boundaries of Computing Across Complex Organizations", in R. Boland and R. Hirschheim, eds., *Critical Issues in Information Systems* (New York, NY: John Wiley & Sons, 1987).

Kuhn, T. S. (1957), *The Copernican Revolution* (Cambridge, MA: Harvard University Press, 1957).

Kuhn, T. S. (1970), *The Structure of Scientific Revolutions*, 2nd ed. (Chicago, IL: University of Chicago Press, 1970).

Lakatos, I. (1976), *Proofs and Refutations* (Cambridge, UK: Cambridge University Press, 1976).

Lakatos, I. and A. Musgrave, eds. (1970), *Criticism and the Growth of Knowledge* (Cambridge, UK: Cambridge University Press, 1970).

Levesque, H. (1986a), "Making Believers out of Computers", *Artificial Intelligence* 30 (1986), pp. 81-108.

Levesque, H. (1986b), "Knowledge Representation and Reasoning", *Annual Review of Computer Science* 1 (1986), pp. 255-287.

Levesque, H. (1988), "Logic and the Complexity of Reasoning", *Journal of Philosophical Logic* 17 (1988), pp. 355-389.

Lewis, D. (1969), *Convention: A Philosophical Study* (Cambridge, MA: Harvard University Press, 1969).

Lewis, D. (1973), *Counterfactuals* (Cambridge, MA: Harvard University Press, 1973).

Lewis, D. (1979), "Scorekeeping in a Language Game", *Journal of Philosophical Logic* 8 (1979), pp. 339-359.

Linsky, L. (1977), *Names and Descriptions* (Chicago, IL: University of Chicago Press, 1977).

Lycan, W. (1984), *Logical Form in Natural Language* (Cambridge, MA: MIT Press, 1984).
Lynd, R. S. (1948), *Knowledge for What?* (Princeton, NJ: Princeton University Press, 1948).
Maida, A. and S. Shapiro (1982), "Intensional Concepts in Propositional Semantic Networks", *Cognitive Science* 6 (1982), pp. 291-330.
Maloney, J. C. (1988), "In Praise of Narrow Minds: The Frame Problem", in J. Fetzer, ed., *Aspects of Artificial Intelligence* (Dordrecht, The Netherlands: Kluwer Academic Publishers, 1988), pp. 55-80.
Marcotty, M. and H. Ledgard (1986), *Programming Language Landscape: Syntax/Semantics/Implementation* (Chicago, IL: Science Research Associates, 1986).
Martins, J. (1989), "Belief Revision", in S. C. Shapiro, ed., *Encyclopedia of Artificial Intelligence* (New York, NY: John Wiley and Sons, 1987), pp. 58-62.
Martins, J. and S. C. Shapiro (1988), "A Model for Belief Revision", *Artificial Intelligence* 35 (1988), pp. 25-79.
Mauer, W. D. (1979), Letter to the Editor, *Communications of the ACM* 22 (1979), pp. 625-629.
McCalla, G. and N. Cercone (1983), "Guest Editors' Introduction: Approaches to Knowledge Representation", *COMPUTER* (October 1983), pp. 12-18.
McCarthy, J. (1968), "Programs with Common Sense", in M. Minsky, ed., *Semantic Information Processing* (Cambridge, MA: The MIT Press, 1968), pp. 403-418.
McCarthy, J. (1987), "Generality in Artificial Intelligence", *Communications of the ACM* 30 (1987), pp. 1030-1035.
McDermott, D. (1981a), "Artificial Intelligence Meets Natural Stupidity", in J. Haugeland, ed., *Mind Design* (Cambridge, MA: MIT Press, 1981), pp. 143-160.
McDermott, D. (1981b), "R1: The Formative Years", *AI Magazine* 2 (1981), pp. 21-29.
Michalos, A. (1969), *Principles of Logic* (Englewood Cliffs, NJ: Prentice-Hall, 1969).
Michalos, A. (1973), "Rationality Between the Maximizers and the Satisficers" *Policy Sciences* 4 (1973), pp. 229-244.
Michie, D. (1985), "Machine Learning and Knowledge Acquisition", *The International Handbook of Information Technology and Automated Office Systems* (Amsterdam, North-Holland: Elsevier Science Publishers, 1985).
Michie, D. (1986), "The Superarticulacy Phenomenon in the Context of Software Manufacture", *Proceedings of the Royal Society* A 405 (1986), pp. 189-212.
Michie, D. and I. Bratko (1986), *Expert Systems: Automating Knowledge Acquisition* (Reading, MA: Addison-Wesley Publishing Company, 1986).
Minsky, M. (1975), "A Framework for Representing Knowledge", reprinted in J. Haugeland, ed., *Mind Design* (Cambridge, MA: The MIT Press, 1981), pp. 95-128.
Minsky, M. (1982), "Why People Think Computers Can't", *AI Magazine* (Fall 1982), pp. 3-15.
Moor, J. (1988), "The Pseudorealization Fallacy and the Chinese Room", in J. Fetzer, ed., *Aspects of Artificial Intelligence* (Dordrecht, The Netherlands: Kluwer Acadmic Publishers, 1988), pp. 35-53.
Morris, C. (1938), *Foundations of the Theory of Signs* (Chicago, IL: University of Chicago Press, 1938).
Newell, A. (1973), "Artificial Intelligence and the Concept of Mind", in R. Schank and K. Colby, eds., *Computer Models of Thought and Language* (San Francisco, CA: W. H. Freeman and Company, 1973), pp. 1-60.

Newell, A. (1981), "The Knowledge Level", *AI Magazine* 2 (1981), pp. 1-20.

Newell, A. and H. Simon (1976), "Computer Science as Empirical Inquiry: Symbols and Search", reprinted in J. Haugeland, ed., *Mind Design* (Cambridge, MA: MIT Press, 1981), pp. 35-66.

Nii, H. P. et al. (1982), "Signal-to-Symbol Transformation: HASP/SIAP Case Study", *AI Magazine* 3 (1982), pp. 23-35.

Nilsson, N. (1980), *Principles of Artificial Intelligence* (Palo Alto, CA: Tioga Publishing Company, 1980).

Nilsson, N. (1983), "Artificial Intelligence Prepares for 2001", *AI Magazine* 4 (1983), pp. 7-14.

Nirenburg, S., ed. (1987), *Machine Translation: Theoretical and Methodological Issues* (Cambridge, UK: Cambridge University Press, 1987).

Nute, D. (1975), "Counterfactuals and the Similarity of Worlds", *Journal of Philosophy* 72 (1975), pp. 773-778.

Nute, D. (1980), "Conversational Scorekeeping and Conditionals", *Journal of Philosophical Logic* 9 (1980), pp. 153-166.

Nute, D. (1988), "Defeasible Reasoning: A Philosophical Analysis in *Prolog*", in J. Fetzer, ed., *Aspects of Artificial Intelligence* (Dordrecht, The Netherlands: Kluwer Academic Publishers, 1988), pp. 251-288.

Pagels, H. (1982), *The Cosmic Code* (New York, NY: Simon & Schuster, 1982).

Pearl, J. (1984), *Heuristics: Intelligent Search Strategies for Computer Problem Solving* (Reading, MA: Addison-Wesley, 1984).

Popper, K. R. (1965), *Conjectures and Refutations* (New York, NY: Harper & Row, 1965).

Popper, K. R. (1972), *Objective Knowledge* (Oxford, UK: The Clarendon Press, 1972).

Popper, K. R. (1978), "Natural Selection and the Emergence of Mind", *Dialectica* 32 (1978), pp. 339-355.

Popper, K. R. (1982), "The Place of Mind in Nature", in R. Elvee, ed., *Mind in Nature* (San Francisco, CA: Harper & Row, Publishers, 1982), pp. 31-59.

Prial, F. (1988), "Secondhand Rose", *The New York Times Magazine* (August 7, 1988), p. 54.

Prior, A. N. (1967a), "Logic, Traditional", in P. Edwards, ed., *The Encyclopedia of Philosophy*, Vol. 5 (New York, NY: Macmillan Publishing Company, 1967), pp. 34-45.

Prior, A. N. (1967b), "Logic, Modal", in P. Edwards, ed., *The Encyclopedia of Philosophy* (New York, NY: Macmillan Publishing Company, 1967), pp. 5-12.

Putnam, H. (1960), "Minds and Machines", in S. Hook, ed., *Dimensions of Mind* (New York, NY: New York University Press, 1960), pp. 138-164.

Putnam, H. (1975), "The Meaning of 'Meaning'", in K. Gunderson, ed., *Language, Mind and Knowledge* (Minneapolis, MN: University of Minnesota Press, 1975), pp. 131-193.

Pylyshyn, Z. (1984), *Computation and Cognition: Toward a Foundation for Cognitive Science* (Cambridge, MA: MIT Press, 1984).

Pylyshyn, Z., ed. (1986), *The Robot's Dilemma: The Frame Problem in Artificial Intelligence* (Norwood, NJ: Ablex, 1986).

Quillian, R. (1967), "Word Concepts: A Theory and Simulation of Some Basic Semantic Capabilities", *Behavioral Science* 12 (1967), pp. 410-430.

Quine, W. V. O. (1951), *Mathematical Logic* (New York, NY: Harper & Row, 1951).

Quine, W. V. O. (1953), "Two Dogmas of Empiricism", *From a Logical Point of View* (Cambridge, MA: Harvard University Press, 1953), pp. 20-46.
Quine, W. V. O. (1960), *Word and Object* (Cambridge, MA: MIT Press, 1960).
Quine, W. V. O. (1975), "Mind and Verbal Dispositions", in S. Guttenplan, ed., *Mind and Language* (Oxford, UK: Oxford University Press, 1975), pp. 83 95.
Quinlan, J. R. (1983), "Learning Efficient Classification Procedures and their Application to Chess End Games", in R. S. Michalski et al., eds., *Machine Learning: An Artificial Intelligence Approach* (Palo Alto, CA: Tioga Press, 1983), pp. 463-482.
Rankin, T. (1988), "When is Reasoning Nonmonotonic?", in J. Fetzer, ed., *Aspects of Artificial Intelligence* (Dordrecht, The Netherlands: Kluwer Academic Publishers, 1988), pp. 289-308.
Rankin, T. (1989), "AI and Conditionality", *IBM AI Reports* 1 (1989), pp. 69-73.
Rapaport, W. J. (1984), "Critical Thinking and Cognitive Development", *American Philosophical Association Newsletter on Pre-College Instruction in Philosophy* 1 (Spring/Summer 1984), pp. 4-5; reprinted in *Proceedings of the American Philosophical Association* 57 (1984), pp. 610-615.
Rapaport, W. J. (1985), "Meinongian Semantics for Propositional Semantic Networks", *Proceedings of the 23rd Annual Meeting of the Association for Computational Linguistics* (Morristown, NJ: Association for Computatonal Linguistics, 1985), pp. 43-48.
Rapaport, W. J. (1986), "Logical Foundations for Belief Representation", *Cognitive Science* 10 (1986), pp. 371-422.
Rapaport, W. J. (1988), "Syntactic Semantics: Foundations of Computational Natural-Language Understanding", in J. Fetzer, ed., *Aspects of Artificial Intelligence* (Dordrecht, The Netherlands: Kluwer, 1988), pp. 81-131.
Rich, E. (1983), *Artificial Intelligence* (New York, NY: McGraw-Hill, 1983).
Romanycia, M. H. and F. J. Pelletier (1985), "What is a Heuristic?", *Computational Intelligence* 1 (1985), pp. 47-58.
Ross, W. D. (1956), *Aristotle* (New York, NY: Meridian Publishers, 1956).
Rudner, R. (1966), *The Philosophy of Social Science* (Englewood Cliffs, NJ: Prentice-Hall, 1966).
Rumelhart, D. E., J. L. McCelland, and the PDP Research Group (1987), *Parallel Distributed Processing* (Cambridge, MA: MIT Press, 1987).
Russell, B. (1905), "On Denoting", reprinted in R. Marsh, ed., *Bertrand Russell: Logic and Knowledge* (New York, NY: Capricorn Books, 1971), pp. 41-56.
Russell, B. (1919), *Introduction to Mathematical Philosophy* (New York, NY: Simon and Schuster, 1919).
Russell, B. (1959), *The Problems of Philosophy* (New York, NY: Oxford University Press, 1959).
Ryle, G. (1949), *The Concept of Mind* (London, UK: Hutchinson, 1949).
Salmon, W. C. (1971), *Statistical Explanation and Statistical Relevance* (Pittsburgh, PA: University of Pittsburgh Press, 1971).
Salmon, W. C. (1984), *Scientific Explanation and the Causal Structure of the World* (Princeton, NJ: Princeton University Press, 1984).
Schank, R. (1973), "Identification and Conceptualizations Underlying Natural Language", in R. Schank and K. Colby, eds., *Computer Models of Thought and Language* (San Francisco, CA: W. H. Freeman, 1973), pp. 187-247.

Schank, R. (1975), *Conceptual Information Processing* (Amsterdam, Holland: North-Holland/American Elsevier, 1975).
Schank, R. (1982a), *Reading and Understanding* (Hillsdale, NJ: Lawrence Erlbaum Associates, 1982).
Schank, R. (1982b), *Dynamic Memory: A Theory of Learning in Computers and People* (Cambridge, UK: Cambridge University Press,1982).
Schank, R. (1983), "The Current State of AI: One Man's Opinion", *AI Magazine (Winter/Spring 1983), pp. 3-8.*
Schank, R. (1984), *The Cognitive Computer* (Reading, MA: Addison-Wesley, 1984).
Schank, R. and R. Abelson (1977), *Scripts, Plans, Goals, and Understanding* (Hillsdale, NJ: Lawrence Erlbaum Associates, 1977).
Schank, R. and J. Carbonell (1979), "Re: The Gettysburg Address: Representing Social and Political Acts", in N. Findler, ed., *Associative Networks: Representation and Use of Knowledge by Computers* (New York, NY: Academic Press, 1979), pp. 327-362.
Schank, R. and C. Riesbeck, ed. (1981), *Inside Computer Understanding: Five Programs Plus Miniatures* (Hillsdale, NJ: Lawrence Erlbaum Associates, 1981).
Shastri, L. (1988), *Semantic Networks: An Evidential Realization and its Connectionist Realization* (Los Altos, CA: Morgan Kaufmann, 1988).
Scheffler, I. (1965), *Conditions of Knowledge* (Glenview, IL: Scott, Foresman and Company, 1965).
Schwartz, S., ed. (1977), *Naming, Necessity, and Natural Kinds* (Ithaca, NY: Cornell University Press, 1977).
Searle, J. (1969), *Speech Acts* (Cambridge, UK: Cambridge University Press, 1969).
Searle, J. (1980), "Minds, Brains, and Programs", *Behavioral and Brain Science* 3 (1980), pp. 417-57; reprinted in J. Haugeland, ed., *Mind Design* (Cambridge, MA: The MIT Press, 1981), pp. 282-306.
Searle, J. (1984), *Minds, Brains, and Science* (Cambridge, MA: Harvard University Press).
Shapiro, S. (1979), "The NSePS Semantic Network Processing System", in N. V. Findler, ed., *Associative Networks: The Representation and Use of Knowledge by Computers* (New York, NY: Academic Press, 1979), pp. 179-203.
Shapiro, S. and W. Rapaport (1987), "SNePS Considered as a Fully Intensional Propositional Semantic Network", in N. Cercone and G. McCalla, eds., *The Knowledge Frontier: Essays in the Representation of Knowledge* (New York: Springer-Verlag, 1987), pp. 262-315.
Shapiro, S. and W. Rapaport (1988), "Models and Minds: A Reply to Barnden", *Northeast Artificial Intelligence Consortium Technical Report TR-8737* (Syracuse, NY: Syracuse University, 1988).
Shope, R. (1983), *The Analysis of Knowing: A Decade of Research* (Princeton, NJ: Princeton University Press, 1983).
Skinner, B. F. (1953), *Science and Human Behavior* (New York, NY: Macmillan Company, 1953).
Skinner, B. F. (1957), *Verbal Behavior* (New York, NY: Appleton-Century-Crofts, Inc., 1957).
Skinner, B. F. (1972), *Contingencies of Reinforcement* (New York, NY: Appleton-Century-Crofts, Inc., 1972).

Slocum, J., ed. (1985), "Special Issue on Machine Translation", *Computational Linguistics* 11/1 and 11/2-3 (1985).

Smart, J. (1984), "Ockham's Razor", in J. Fetzer, ed., *Principles of Philosophical Reasoning* (Totowa, NJ: Rowman and Allanheld, 1984), pp. 118-128.

Smolensky, P. (1988), "On the Proper Treatment of Connectionism", *Behavioral and Brain Sciences* 11 (1988), pp. 1-23.

Sneed, J. (1971), *The Logical Structure of Mathematical Physics* (Dordrecht, Holland: D. Reidel Publishing Company, 1971).

Stalnaker, R. (1968), "A Theory of Conditionals", *American Philosophical Quarterly*, Supplementary Monograph No. 2 (Oxford, UK: Basil Blackwell, 1968), pp. 98-112.

Stich, S., ed. (1975), *Innate Ideas* (Berkeley, CA: University of California Press, 1975).

Stich, S. (1983), *From Folk Psychology to Cognitive Science* (Cambridge, MA: MIT Press, 1983).

Stillings, N., et al. (1987), *Cognitive Science: An Introduction* (Cambridge, MA: MIT Press, 1987).

Sumner, L. and J. Woods, eds. (1969), *Necessary Truth: A Book of Readings* (New York, NY: Random House Publishers, 1969).

Suppe, F., ed. (1977), *The Structure of Scientific Theories*, 2nd ed. (Urbana, IL: University of Illinois Press, 1977).

Suppes, P. (1967), *Set-Theoretical Structures in Science* (Stanford, CA: mimeographed, 1967).

Tarski, A. (1956), "The Concept of Truth in Formalized Languages", reprinted in A. Tarski, *Logic, Semantics, and Metamathematics* (Oxford, UK: Oxford University Press, 1956), pp. 152-278.

Teller, P. (1980), "Computer Proof", *Journal of Philosophy* 77 (1980), pp. 797-903.

Turing, A. (1950), "Computing Machinery and Intelligence", reprinted in A. Anderson, ed., *Minds and Machines* (Englewood Cliffs, NJ: Prentice-Hall, 1964), pp. 4-30.

Twain, M. (1875), "The Jumping Frog [In English. Then in French. Then Clawed Back into a Civilized Language One More by Patient, Unremunerated Toil.]", in J. Kaplan, ed., *Great Short Works of Mark Twain* (New York, NY: Harper & Row, 1967), pp. 79-95.

Tymoczko, T. (1979), "The Four Color Theorem and Its Philosophical Significance", *Journal of Philosophy* 76 (1979), pp. 57-83.

Van den Bos, J. (1979), Letter to the Editor, *Communications of the ACM* 22 (1979), p. 623.

Vaughan, R. (1988), "Maintaining an Inductive Data Base", in J. Fetzer, ed., *Aspects of Artificial Intelligence* (Dordrecht, The Netherlands: Kluwer Academic Publishers, 1988), pp. 323-335.

Walsh, W. H. (1967), "Kant, Immanuel", in P. Edwards, ed., *The Encyclopedia of Philosophy*, Vol. 4 (New York, NY: The Macmillan Publishing Company, 1967), pp. 305-324.

Waterman, D. A. (1986), *A Guide to Expert Systems* (Reading, MA: Addison- Wesley Publishing Company, 1986).

Webster's (1988), *Webster's New World Dictionary of the American Language*, 3rd college ed. (New York, NY: Simon & Schuster, 1988).

Wilensky, R. (1981), "PAM", in R. Schank and C. Reisbeck, eds., *Inside Computer Understanding: Five Programs Plus Miniatures* (Hillsdale, NJ: Lawrence Erlbaum Associates, 1981), pp. 136-179.

Wilensky, R. (1984), *LISPcraft* (New York, NY: W. W. Norton & Company, 1984).

Winograd, T. (1983), *Language as a Cognitive Process*, Vol. I (Reading, MA: Addison-Wesley, 1983).

Winograd, T. and F. Flores (1986), *Understanding Computers and Cognition* (Norwood, NJ: Ablex Publishing, 1986).

Winston, P. (1984), *Artificial Intelligence*, 2nd ed. (Reading, MA: Addison-Wesley, 1984).

Woods, W. A. (1975), "What's in a Link: Foundations of Semantic Networks", in D. Bobrow and A. Collins, eds., *Representation and Understanding* (New York, NY: The Academic Press, 1975), pp. 35-82.

Woods, W. A. (1983), "What's Important about Knowledge Representation?", *COMPUTER* (October 1983), pp. 22-27.

NUMBERED DEFINITIONS

As a convenience, a list of the numbered definitions that occur in the text is provided here. It should be observed that some terms are defined in the text before the appearance of these numbered definitions ("open system" and "closed system", for example, are first introduced in Chapter 2). It should also be observed that the term, "dispositional predicate" as defined by (D2) is a special, restricted sense explored in the pursuit of a more adequate conception. For further references, of course, the "Index of Subjects" is provided.

LIST OF FIGURES

INDEX OF NAMES

INDEX OF SUBJECTS

STUDIES IN COGNITIVE SYSTEMS

Series Editor: James H. Fetzer, *University of Minnesota*

1. J. H. Fetzer (ed.): *Aspects of Artificial Intelligence*. 1988
   ISBN Hb 1-55608-037-9; Pb 1-55608-038-7
2. J. Kulas, J.H. Fetzer and T.L. Rankin (eds.): *Philosophy, Language, and Artificial Intelligence*. 1988 ISBN 1-55608-073-5
3. D.J. Cole, J.H. Fetzer and T.L. Rankin (eds.): *Philosophy, Mind and Cognitive Inquiry*. Resources for Understanding Mental Processes. 1990
   ISBN 0-7923-0427-6
4. J.H. Fetzer: *Artificial Intelligence*. Its Scope and Limits. 1990
   ISBN Hb 0-7923-0505-1; Pb 0-7923-0548-5
5. H.E. Kyburg, Jr., R.P. Loui and G.N. Carlson (eds.): *Knowledge Representation and Defeasible Reasoning*. 1990 ISBN 0-7923-0677-5

KLUWER ACADEMIC PUBLISHERS

9 780792 305057